Synchrotron Radiation in Chemistry and Biology II

Synchrotron Radiation Studies on Insect Flight Muscle
K. J. V. Poole, G. Rapp, Y. Maéda and R. S. Goody

Protein Single Crystal Diffraction
I. D. Glover, J. R. Helliwell and M. Z. Papiz

Synchrotron Light on Ribosomes: The Development of Crystallographic Studies of Bacterial Ribosomal Particles
K. S. Bartels, G. Weber, S. Weinstein, H.-G. Wittmann and A. Yonath

Application of EXAFS to Biochemical Systems
S. S. Hasnain

Structure, Dynamics and Growth Mechanisms of Metal-Metal and Metal-Semiconductor Interfaces by Means of SEXAFS
D. Chandesris, P. Roubin and G. Rossi

A Storage Phosphor Detector (Imaging Plate) and Its Application to Diffraction Studies Using Synchrotron Radiation
Y. Amemiya, Y. Satow, T. Matsushita, J. Chikawa, K. Wakabayashi and J. Miyahara

Photoacoustic X-ray Absorption Spectroscopy
T. Masujima

AKADEMIE-VERLAG BERLIN

Topics in Current Chemistry · Vol. 147

Synchrotron Radiation in Chemistry and Biology II

Synchrotron Radiation in Chemistry and Biology II

With 91 Figures and 10 Tables

AKADEMIE-VERLAG BERLIN
1988

Die Originalausgabe erscheint im Springer-Verlag
Berlin · Heidelberg · New York · London · Paris · Tokyo
in der Schriftenreihe "Topics in Current Chemistry",
Volume 147

Vertriebsrechte für die sozialistischen Länder:
Akademie-Verlag Berlin

Vertriebsrechte für alle Staaten mit Ausnahme der sozialistischen Länder:
Springer-Verlag Berlin · Heidelberg · New York · London · Paris · Tokyo

ISBN 3-540-19040-6 Springer-Verlag Berlin · Heidelberg · New York
ISBN 0-387-19040-6 Springer-Verlag New York · Heidelberg · Berlin

ISBN 3-05-500560-0 Akademie-Verlag Berlin

Erschienen im Akademie-Verlag Berlin, DDR-1086 Berlin, Leipziger Straße 3—4
Lizenznummer: 202 · 100/527/88
Printed in the German Democratic Republic
Gesamtherstellung: VEB Druckerei „Thomas Müntzer", 5820 Bad Langensalza
LSV 1275
Bestellnummer: 763 875 7 (3072/147)
12800

Preface

This is the second of three volumes in the Springer series "Topics in Current Chemistry" devoted to applications of synchrotron radiation in chemistry and biology. It contains contributions dealing with X-ray diffraction, spectroscopy, and technical developments. The article by K. Poole et al. describes structural and time-resolved studies on muscle fiber and represents a continuation of the early applications of synchrotron radiation in biology, started at DESY by K. Holmes, G. Rosenbaum, and coworkers. The contributions by I. Glover et al. and K. Bartels et al. deal with protein crystallography to obtain high resolution structures. Although this method antedates synchrotron radiation laboratories, its scope has expanded dramatically because of the new X-ray sources and other technical innovations. S. Hasnain and D. Chandesris et al. give examples of spectroscopic applications in biology and chemistry. Progress in these areas depended not only advances in experimental methods but also in improvements of data analysis and theory. Finally, the articles by Y. Amemiya et al. and T. Masujima illustrate novel technical developments with unexpected potential when combined with synchrotron radiation. The articles collected in this volume describe results obtained at synchrotron radiation laboratories in France, Britain, Germany, Japan, and the United States, reflecting the international nature of this research and the countries actively engaged in it. It is hoped that this volume serves as a guide for scientists considering applying synchrotron radiation to their own research problems.

Hamburg, February 1988 Eckhard Mandelkow

Table of Contents

Table of Contents of Volume 145

Synchrotron Radiation Studies on Insect Flight Muscle

Katrina J. V. Poole, Gert Rapp, Ynichiro Maeda* and Roger S. Goody

Department of Biophysics, Max-Planck Institut für Medizinische Forschung, 6900 Heidelberg, FRG;

Table of Contents

---

* EMBL Outstation, Hamburg, FRG

W

Insect flight muscle has proved particularly suitable for the structural investigation of the contractile proteins and their interactions. Here we review the low angle x-ray scattering experiments performed on this muscle using synchrotron radiation as a source. We briefly outline the present concensus on the cross-bridge mechanism of muscle contraction, and introduce insect muscle structure and muscle fibre diffraction. Details of x-ray beamline and camera configurations are considered with particular reference to the requirements necessary for small specimen work. A brief summary of the early experimental work on equilibrium or pseudoequilibrium states, such as that trapped by the ATP analogue AMP-PNP, is followed by a more detailed account of recent time-resolved experiments. These include measurements of the intensities of particular low-angle reflections during stretch activation and following the rapid, photolytic release of ATP from caged-ATP.

1 Introduction

Insect flight muscle has played an important role in the development of synchrotron radiation as a source for x-ray scattering experiments on biological specimens. This is because the group which pioneered the development was interested in the structure and function of this muscle, and the first low angle scattering data from a biological specimen (actually the first x-ray scattering data from a specimen of any kind) to be obtained using synchrotron radiation were from the flight muscle of the tropical water bug *Lethocerus maximus* (Rosenbaum et al., 1971). Studies on this muscle continued in the ensuing years, but it has taken 15 years to begin to achieve the original aim of obtaining dynamic structural data in the millisecond time range.

2 Muscle Contraction

There are a variety of systems existing within the animal kingdom in which motion or force is generated, but the contraction of striated muscle cells has received by far the most attention, since the two basic contractile proteins, actin and myosin, are present in large amounts within these cells and are organised into a highly ordered filament array. These features have therefore allowed the biochemical characterisation of these proteins and their interactions in solution as well as their structural characterisation within the cell using averaging techniques such as x-ray diffraction. A striated muscle fibre consists of a bundle of sub-fibres called myofibrils, each of which is further divided into a longitudinal series of sarcomeres resulting in the characteristic striped appearance. The sarcomere, shown in Fig. 1a, is the basic contractile unit and consists of a double array of interdigitating filamentous aggregates of actin and myosin which slide past each other during muscle shortening. Consequently, the central problem in muscle research has been to determine the origin of the shear force generated between these filaments.

The thin filament has the simpler structure and consists of a double helical arrangement of globular actin monomers (42 kD molecular weight, M.W.) which has an axial repeat of 77.0 nm and a subunit repeat of 5.9 nm. In most striated muscles the regulatory proteins, troponin and tropomyosin are associated with this filament and act, via calcium binding, as an on/off switch for contraction. The thicker of the two filaments (backbone diameter ca. 20 nm in insect flight muscle c.f. <10 nm for the actin filament) are aggregates of myosin molecules which posess two heavy chains (M.W. 200 kD) and four light chains (M.W. ca. 20 kD), as shown in Fig. 1b. The heavy chains have a globular head (M.W. 120 kD) and an α-helical region which interacts with that of the partner chain to form a coiled-coil rod, and it is these structures which aggregate to form the thick filament backbone, leaving the heads protruding on a 40 nm flexible arm. The packing of the rods and the arrangement of the heads around the filament varies considerably between different muscle types (Wray et al., 1975; Wray, 1979; Squire, 1981), but the surface lattice of the cross-bridge projections appears to be helical. The repeat of this helix in insect flight muscle is thought to be 115.5 nm since the 14.5 nm, 23.5 nm, and 38.5 nm periodicities apparent in longitudinal electron micrographs (Reedy et al., 1965; Reedy, 1967; Reedy and Garrett, 1977) are all orders of this repeat.

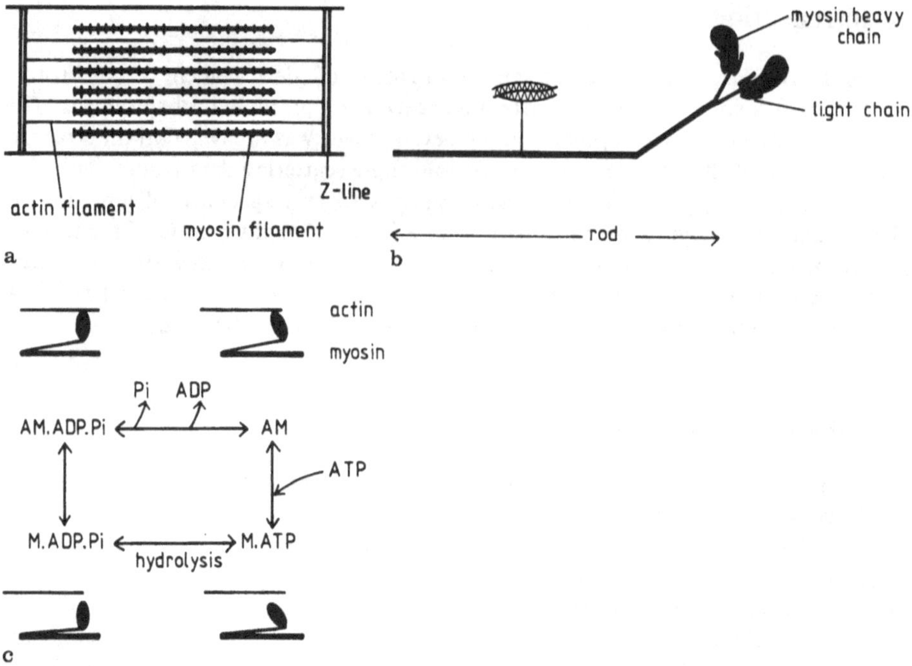

Fig. 1a–c. a The sarcomere, consisting of the thick (myosin) filaments which have radial projections, and the thin (actin) filaments. Overlap between the two increases as the muscle shortens. In insect flight muscle this structure is ca. 2.5 μm long, and the thick filaments are ca. 2.2 μm long. **b** The myosin molecule (Mol. wt. ca. 480 kDalton). The heavy chains have a globular head plus coiled-coil rod. The 4 light chains may be involved in regulating the interaction of the head with the thin filament in insect muscle (Lehman, 1977). The rod is 140 nm long and the 100 nm distal portions aggregate to form the stable, heliçally ordered backbone of the thick filaments. **c** The cross-bridge cycle. Reactions are reversible but there is a net flux in the clockwise direction. There is a fairly large increase in free energy when ATP binds to actomyosin (AM) which causes cross-bridge dissociation. Following the hydrolytic step, the bridge reattaches and the consequent loss of products is accompanied by a large decrease in free energy which drives the structural change which causes the filaments to slide. (After Lymn & Taylor, 1971)

The head part of myosin has received most attention since it posesses both an enzymatic site for ATP hydrolysis and an actin binding site, and the cross-bridge theory of muscle contraction postulates that the energy derived from ATP cleavage drives a cyclical interaction of the heads with the actin filaments which are pulled along by a configurational change in the attached myosin bridge (A. F. Huxley, 1957; H. E. Huxley, 1969; Huxley & Kress, 1985; Eisenberg, 1986; Hibberd and Trentham, 1986). Unfortunately the detailed structure of the head is not known, although its recent crystallisation by Winkelman et al. (1985) should remedy this situation in the future. Considerable progress has, however, been made in elucidating the kinetics and energetics of ATP hydrolysis by actin and myosin in solution (see reviews by Taylor, 1979; Eisenberg, 1986; Kodama, 1985) and Fig. 1c shows a simplified scheme where the cross-bridge diagrams represent possible structural counterparts of the different states. Basically, the nature of the magnesium nucleotide complex bound at the active site dramatically influences the affinity of myosin for actin, the ATP and ADP.Pi states having a very low affinity ($K_d = 10^{-5}$ M), and the ADP and nucleotide free heads an affinity more than 10^5 fold higher. It is thought that the energy of ATP cleavage is stored in a metastable state of the bridge, the M.ADP.Pi state which is dissociated from the thin filament, and is released as a conformational change, the power stroke, probably when the hydrolysis product, Pi, is released from the actin bound bridge.

The determination of the nature and precise location of the structural change that forms the power stroke is problematical since the asynchronous cycling of bridges required to produce steady force or shortening prevents the application of structural averaging techniques. This problem has therefore been approached in two different ways. Either one attempts to trap the crossbridges in particular states within the cycle for equilibrium or pseudo-equilibrium measurements, or one tries to temporarily synchronize all the bridges by imposing rapid mechanical or biochemical perturbations on muscle fibres in order to measure transient structural changes. The first of these approaches has been used extensively in the investigation of insect muscle structure, and the non-physiological, ATP-free state or rigor state has received the most attention since it is easy to achieve and is a possible candidate structure for the end of the power stroke (Miller & Tregear, 1972; Holmes et al., 1980). Alternatively, a number of structural studies have used analogues of ATP in order to mimic other states of the attached bridge, and some of these are described below. More recently the second approach has been applied to the insect fibre system and in the last section of this paper we describe some time resolved diffraction measurements made following rapid activation and rapid perturbations in nucleotide concentration.

3 Insect Flight Muscle as the Specimen

Some insect flight muscles have become highly specialised for a high frequency oscillatory mode of operation. In these, so-called fibrillar muscles, the membrane systems associated with neuromuscular excitation and calcium pumping are much reduced in favour of a mechanical triggering mechanism. The myofibrils are more plentiful and are particularly well ordered within the cell, as are the filaments within the sarcomere, exemplified by the cross section shown in Fig. 2d. Unlike the situation in other striated muscles, the filaments overlap almost completely and the sarcomere shortens by only a few percent of its length during oscillatory contraction. Also, the thick filaments appear to be connected to the ends of the sarcomere, thus providing longitudinal structural continuity. Consequently, the myofibrils and hence whole fibres are structurally very stable, even after demembranation, which conveniently allows the study of the isolated contractile machinery. The properties and behaviour of these muscles have been reviewed (Pringle, 1977; Tregear, 1977, 1981 and 1983). The dorsal longitudinal flight muscles of giant Belostomatid water bugs of the genus *Lethocerus* have become the standard preparation since they reach over 1 cm in length and are therefore very easy to handle. They are usually skinned in a buffered glycerol-salt solution and can be stored for many months at temperatures of −20 °C before use.

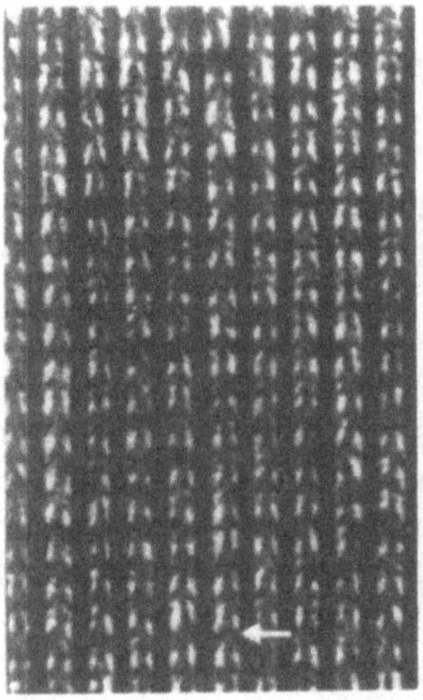

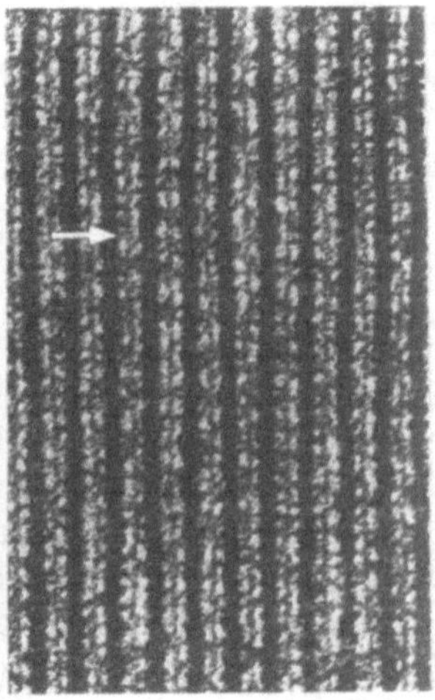

b

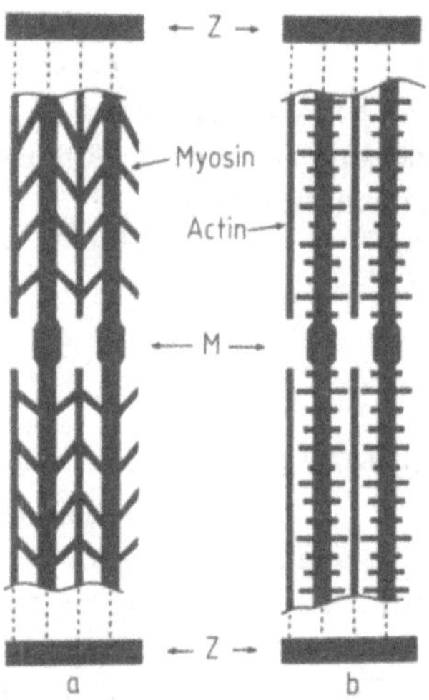

c

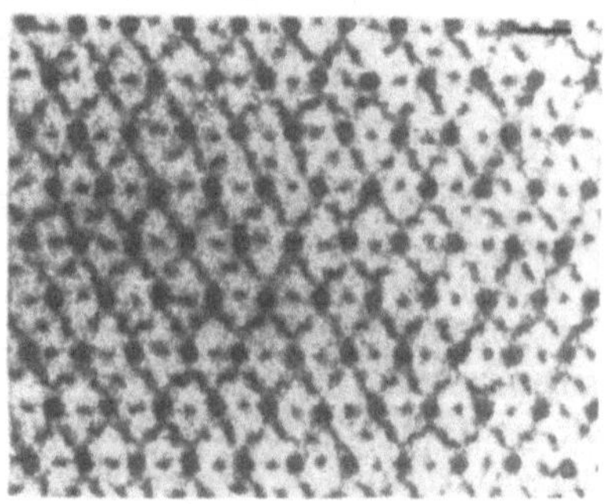

d

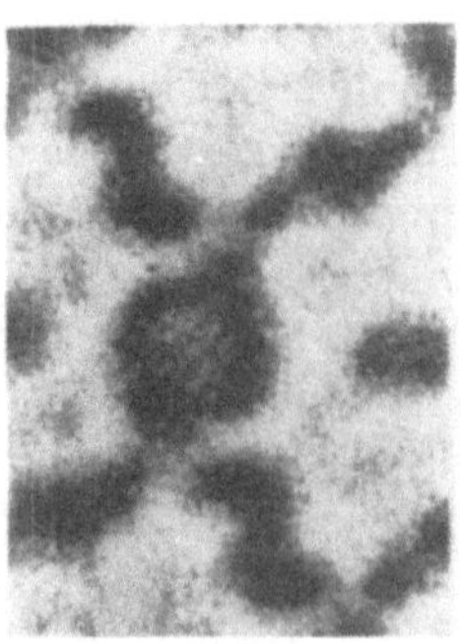

e

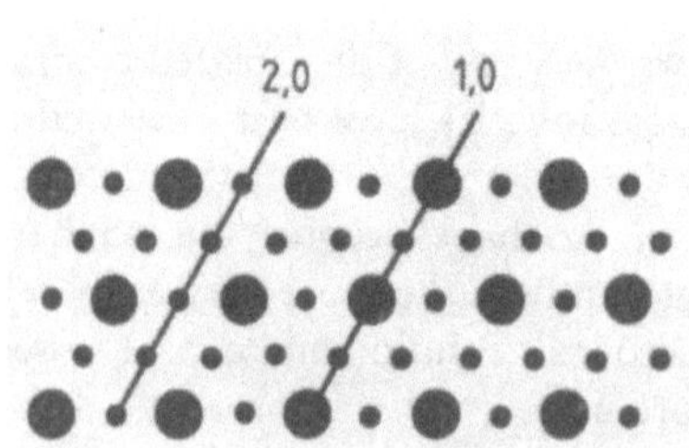

f

Over the last twenty years biophysical work on this preparation has concentrated mainly on the elucidation of the filament structure and cross-bridge conformations (Reedy et al., 1965; Squire et al., 1977; Wray, 1979; Clarke et al., 1986; Reedy et al., 1987), and on the mechanical characterisation of various equilibrium states and of the kinetics of the cross-bridge cycle (Jewell & Rüegg, 1966; White, 1970; Tregear, 1977; Güth et al., 1981; White & Thorson, 1983). The biochemistry of ATP hydrolysis by the insect proteins has received less attention than that of vertebrate muscle proteins, primarily because of shortage of tissue, but recently aspects of the biochemical kinetics have been investigated (White et al., 1986).

The electron micrographs of Fig. 2, taken from very thin sections of *Lethocerus* muscle, illustrate the state of the art in the imaging of cross-bridge structures. The longitudinal sections show that bridging structures are not clearly seen in the presence of ATP, the mass of the myosin heads appearing in 14.5 nm spaced shelves along the thick filament. In the absence of ATP the bridges are clearly seen attaching to the thin filaments in a chevron pattern, and Fig. 2c illustrates the simplified interpretation of these images in terms of angled rigor attachments and perpendicularly oriented detached bridges. More detail on bridge conformations and their arrangement between the filaments can be obtained from serial cross-sections taken along the myofibril. Figure 2d shows one section which illustrates the projection of cross bridges to four different actin filaments. Optical filtering of this image yields the clarified averaged structure shown in Fig. 2e.

Images such as these and the three-dimensional reconstruction of the structures producing them have made the cross-bridge arrangement in insect muscle the most fully characterised, and have proved invaluable in the interpretation of x-ray diffraction data from this preparation.

4 X-Ray Diffraction by Muscle Fibres

Muscle fibres can be considered as a bundle of microcrystals which are randomly organised around the long axis so that irradiation with a monochromatic beam gives rise to a series of reflections arising from the repeating structures associated with the myofilaments. The meridional reflections, lying along the axis of the fibre diagram parallel to the fibre axis, arise from axial periodicities present within the filaments, and in insect muscle, the 14.5 nm period of the cross-bridge projections from the

---

Fig. 2a–f. The structure of insect flight muscle. **a** is a 20 nm thick, longitudinal section of actomyosin in a rigor myofibril showing the double chevron arrangement of cross-bridging to actin. Chevrons point towards the middle of the sarcomere. The thick filaments show no 14.5 nm striping. **b** is a similar, ATP-relaxed section showing clear 14.5 nm striping together with 90° projections of mass from the thick filaments. **c** shows a diagrammatic interpretation of the crossbridge arrangements seen in the above sections (from Reedy et al., 1965). **d** shows a 11–15 nm cross-section of a rigor myofibril, illustrating the beautiful regularity of the filament lattice and the 'flared-X' arrangement of cross-bridges projecting from the each thick filament to 4 adjacent thin filaments. **e** is an enlargement of an optically filtered image of the flared-X structure demonstrating the fine structural detail obtainable from this specimen. **f** is an illustration of the filament lattice seen in **d** showing the 1010 and 2020 crystal planes. All the electron micrographs are taken from Reedy et al. (1987)

thick filaments in the presence of ATP gives rise to the strongest of these reflections, the first order of which is referred to as the 14.5 nm reflection. The helical nature of the actin and myosin filaments gives rise to a series of horizontal streaks of intensity called layer lines, their axial and radial positions being determined by the parameters characterizing the helices, and their shape and intensity by the shape and positioning of the subunits within them (Holmes & Blow, 1966; Huxley & Brown, 1967; Squire, 1981).

If the molecular filaments had no specific transverse relation to each other then the above information would represent the molecular transform of the filaments alone.

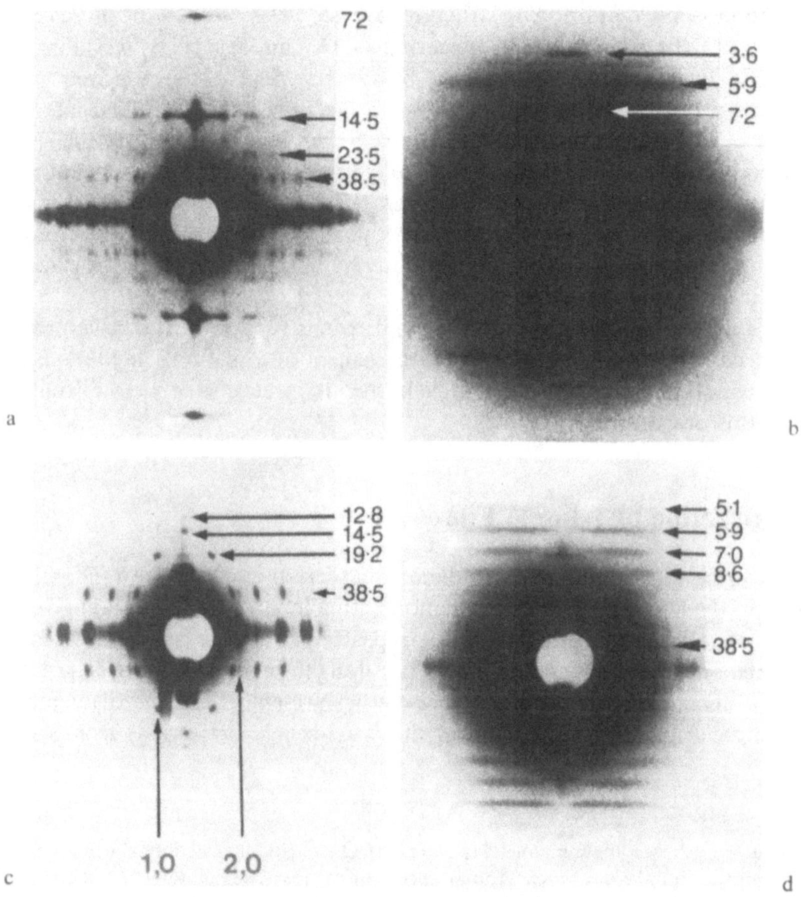

Fig. 3a–d. X-ray diffraction patterns from glycerinated *Lethocerus* fibres in the presence (**a & b**) and absence (**c & d**) of ATP. **a & c** are the lower order layer lines, and B & D are stronger exposures at a 50% shorter camera length showing the outer layer lines. The relaxed pattern is characterised by the strong myosin 14.5 nm and 7.2 nm reflections, and has weak layer line reflections at 38.5 nm and 23.5 nm, associated with the helical organisation of the myosin. The stronger exposure in B indicates the presence of the 5.9 nm actin subunit periodicity. In the rigor patterns, the arrows mark the positions of the much intensified actin layer line system. Exposures were taken on the DESY camera (7.2 GeV, 9 mA, 1 hr for **b**. From Barrington-Leigh et al. (1977)

However, in most striated muscles the filaments are held in an hexagonal lattice, and this leads to a sampling of the molecular transform by the transform of this lattice. The particularly precise filament arrangement in fibrillar muscle, illustrated in cross sections of Fig. 2d, causes a particularly pronounced sampling of the reflections resulting in the series of sharp reflections along the equator and inner layer lines of the patterns shown in Fig. 3a & c. The position of the spots in the equatorial direction is inversely proportional to the spacing between the filaments, and their intensities provide information on the distribution of mass between them. The two innermost of these spots are the strongest reflections in the scattering diagram and are referred to as the 1,0 and the 2,0 reflections since they arise from the 1010 and 2020 crystal planes of the lattice which are illustrated in Fig. 2f. This means that they are also the easiest to measure, and since they show dramatic changes in their relative intensities in response to movements of the myosin heads such as that taking place when bridges bind in rigor (compare the rigor and relaxed equatorial patterns of Fig. 4b), it has been possible to monitor these reflections in a number of time resolved studies as described below.

The photographs of Fig. 3, showing the low angle scattering from insect muscle in the relaxed and rigor states, were taken on the original, small focus DESY camera and illustrate all the fibre diffraction features described above. However, despite the wealth of information available from such pictures, we are far from 'solving' the structures giving rise to them since, unlike crystallographers, we have neither a precise transform of the molecules nor a way of determining the phases of the reflections. A number of attempts to understand the intensity distributions of the major reflections have been made by comparing them with model predictions (Miller & Tregear, 1972; Holmes et al., 1980), but usually, the locations of the reflections are interpreted in terms of known periods within the filament lattice, and the intensities, or the changes in intensity induced by changes in physiological state of the fibres, are interpreted rather simplistically in terms of changes in the amount of mass marking the particular periods. The positions of the major reflections arising from the actin and myosin repeats are indicated on the patterns and it is clear that the binding of crossbridges to actin in the rigor state causes an enhancement of the intensities associated with the helical organisation of the thin filament, and a reduction of those associated with the thick filament geometry.

5 X-Ray Benches

Four different x-ray benches have been used in the work reviewed here. They are described briefly below, and some additional information on two of them is given in the Appendix.

DESY bench: The first optical bench to be used for scattering and diffraction experiments using synchrotron radiation was constructed on the electron synchrotron (Deutsches Elektronen Synchrotron) in Hamburg, F.R.G. in the early 1970's. It consisted of two 20 cm focussing polished quartz mirrors followed by a quartz monochromator (replaced later by germanium to give more intensity) giving 0.15 nm monochromatic radiation. This bench produced a beam with considerably less overall intensity than the storage ring benches constructed later, but had the advantage of having by far the smallest focal size of all four arrangements discussed here (compare the spot sizes in the diffraction diagrams of Figs. 3a & c, taken on this bench, with those of Fig. 4a taken on X33 described below). The bench is described in detail by Barrington-Leigh and Rosenbaum (1976). It was dismantled after the construction of beam lines on the storage ring (DORIS).

X11: This was originally a bench of similar design to the DESY bench but with 8 mirrors instead of two. It was built on the storage ring DORIS and had the capability of variable wavelength (Rosenbaum, 1979; Rosenbaum and Harmsen, 1978). Although the mirrors were designed to be focussed, this option was never realized in practice. Total flux was considerably higher than on the DESY bench, mainly due to the higher current and smaller bending radius of the storage ring in comparison with the DESY synchrotron. A major disadvantage for experiments with insect flight muscle, and for small specimens in general, was the large focal size. This was mainly due to the dimensions and position of the virtual source, but also to lack of focussing in the vertical direction and the limited demagnification (see Appendix for more details). The bench has been dismantled and reconstructed for crystallographic work.

X13: This was an instrument of similar optical design to X11, but lacking the variable wavelength facility. In terms of beam quality it had similar advantages and limitations to X11. The camera was used to introduce a triangular monochrometer (Lemònnier et al., 1978) in place of the old rectangular design. For a detailed desciption see Hendrix et al. (1979).

X33: This bench was constructed on the storage ring DORIS at a different location (HASYLAB). It had the new feature of a triangular monochrometer in front of 8 non-focussing mirrors, which led to a considerable increase in useful lifetime of the mirrors due to much reduced loading. This bench is most wasteful of flux density, since the virtual source is even larger and there is no vertical focussing capability, and its position is even less favourable than for X11 and X13, thus reducing the demagnification achieved. In spite of this, most dynamic experiments have been performed with this camera, mainly due to availability and level of support.

6 Detectors

X-ray film: Early work on insect flight muscle using synchrotron radiation was performed almost exclusively using film as a 2D detector (Goody et al., 1975). Film was particularly useful on the DESY bench, since the small focal size made it highly suitable for use with a detector with high spatial resolution. X11, X13, and particularly X33, are much less suitable for work with x-ray film, since the large focal size means that reflections are much more diffuse and difficult to distinguish from the inherent background ("fog") level present even with the best and freshest film samples. This property of x-ray film, i.e. the fog level, together with the relatively low sensitivity, are its main drawbacks compared with other detectors, its main advantages being the high spatial resolution and ease of use. In addition, film cannot be used, at least not conveniently, for time-resolved measurements. Nevertheless, film has been of importance for characterising static and quasi-static states, and these have formed the basis for later time-resolved measurements.

1-dimensional position sensitive detectors: 1-D electronic detectors of the delay line type were used relatively early with insect flight muscle and synchrotron radiation (Goody et al., 1976) and continue to be of great use (M. K. Reedy et al., 1983; Goody et al., 1985a). However, for time resolved experiments making maximal use of the available intensity, the highest count rates which can be tolerated by this type of counter are too low. Thus for genuine time resolved studies, a multiwire detector (Hendrix et al., 1982) has been used in recent years (Rapp et al., 1986; Poole et al., 1987). In this type of detector, each wire (typically 128) represents 1 channel, and the space between the wires determines the spatial resolution. Since this is of the order of 1 mm, the resolution is considerably worse than for delay line counters (ca. 0.2 mm), and much worse than film. However, this can often be tolerated in the interest of high count rates. The use of such a detector to measure the sharp reflections from insect muscle on an x-ray bench with optical characteristics comparable to the DESY bench would, however, be unsatisfactory in the sense that potential spatial resolution would be lost. Both types of linear detector used offer the advantage of high sensitivity and negligible inherent background.

Electronic 2-dimensional counters: Multiwire counters have not yet found much application in conjunction with insect flight muscle, although some data obtained recently using a detector built by A. Gabriel (see Maéda et al., 1986) are presented later in this article. In general, detectors presently available suffer from a relatively low sensitivity (although this is much higher than film), low maximal count rates and low spatial resolution and so are not particularly suitable for time resolved studies. They do offer the advantages of low inherent background and of producing data in a form suitable for rapid processing. Improvements in any of the areas mentioned above are eagerly awaited, since the advantage to be gained from collecting time-resolved data of complete diffraction patterns is obvious.

A two dimensional television type detector was used in earlier work on X11 for examining static states in insect flight muscle (Tregear et al., 1979). This was an integrating detector with no provision for time resolved measurements due to the long read-out time of the instrument. Despite the high inherent background, which almost prevented collection of the diffuse reflections in the diffraction pattern, the high sensitivity and spatial resolution made this counter useful in the short time it was available.

Fuji plates: A conceptually new x-ray detection system has recently been developed by Fuji Photo Film (Tokyo) and has been used at a synchrotron source (Miyahara et al., 1986). This system employs photostimulatable luminescence of energy (derived from the energy of the x-ray photons) stored temporarily in a $BaFBr:Eu^{2+}$ phosphor. According to our experience (Maéda et al., 1987), the system has several advantages. Spatial resolution is good (at present 0.1 mm, but the plates themselves have an inherent resolution much better than this). The sensitivity at 0.15 nm is about 50x better than for x-ray film, and is almost as good as an electronic detector. There is no upper limit to the count rate, making it particularly suitable for synchrotron radiation. The dynamic range of the phosphor is high, of the order of 10^5, although the dynamic range of the system is limited by that of the photomultiplier tube used for reading out, which is ca. 10^3. The area of the plates is large (20×25 cm), allowing full patterns to be obtained from single exposures. The inherent background is much lower than that of x-ray film as long as the plates are "read" within a few hours of exposure. The system is not easily adapted for use in time-resolved experiments, but it seems to be the best detector yet available for characterising changes between static or pseudo static end states prior to real time resolved studies.

7 Special Requirements for Small Specimen Work

Useable bundles of glycerinated insect flight muscles are small specimens for x-ray diffraction investigations. Although for some experiments (in general static), relatively large numbers of fibres can be used (up to ca. 100; single fibre thickness ca. 50 microns), most experiments, particularly dynamic ones, require much smaller bundles (ca. 20 fibres) or ideally in certain cases single fibres. This means that not only is the amount of material which can scatter x-rays very small, but that the additional problem of directing a reasonable fraction of the available intensity through the sample arises. The latter problem is particularly serious with presently available synchrotron sources and benches due to the relatively large beam sizes. Ironically, but perhaps understandably in view of the interests of the group which set up the first diffraction bench on DESY, the requirement of small beam size was satisfied more nearly by this bench than by any of the subsequent ones. The focal size was ca. 0.3×0.1 mm (horizontal $\times$ vertical) and the size at the specimen was between 5×1 mm and 2×0.3 mm, depending on the specimen-detector distance. Since *Lethocerus maximus* or *indicus* fibres are over 1 cm long, and a bundle of 50 fibres is a few tenths of a millimetre thick, a large fraction of the total flux could be usefully used, and it was not necessary to limit the beam size, which always introduces new potential sources of background. An additional advantage of the small beam size was that high quality static exposures using film could be obtained, and such recordings from this bench were far superior to those from X11, X13 or X33. The problems described are particularly severe with X33, where the very large focal size makes film recording of no practical use. As already mentioned, this problem is less serious with other methods of detection, but the serious problem of the the beam size being much larger than that of the specimen remains for small specimens. X33 has been very satisfactory for measurements on large specimens, such as living frog muscle (e.g. Huxley et al., 1982; Huxley, 1984; Poulsen & Lowey, 1987). These muscles are so large that they can be mounted with the fibre axis vetical and still be larger than the beam in the horizontal direction. This has the considerable advantage that the direction of smaller focal size is in the meridional direction of the diffraction pattern, so that the width of the layer lines is less than with the horizontal arrangement, leading to a better signal to noise ratio. This arrangement would waste most of the available flux with thin specimens such as small bundles of insect flight muscle, or indeed of any other muscle. Thus, there is no substitute for small focal size when dealing with small specimens. Factors affecting the focus size are discussed in more detail in the appendix.

8 Static Experiments on Insect Flight Muscle

The first detailed experiments to be reported which involved the use of synchrotron radiation on insect flight muscle fibres were concerned with an examination of the effects of ATP-analogues on their structure (Goody et al., 1975). This work was done on the DESY bench. High intensity was needed not for temporal resolution of events occurring in the contractile cycle, but to allow a series of low-angle diffraction exposures from different states of the same muscle bundle to be obtained in a reasonably short time (several hours rather than the several days needed using a rotating anode source). The object of these studies was to stop the contractile cycle at different stages, using appropriate nucleotides and analogues, and to characterize these states structurally. Two main results emerged from this work. Using the analogue ATP(γ-S) (Goody and Eckstein, 1971), in which one of the oxygens of the terminal phosphate group is replaced by sulphur, kinetic studies (Bagshaw et al., 1983) had already shown that the cycle is stopped at a state in which myosin heads are dissociated from thin filaments, but in which the analogue is not yet hydrolyzed to give the myosin-products complex characteristically achieved in the presence of ATP (but absence of Ca^{2+}). Despite this difference in the nature of the nucleotide at the active site of myosin, the low-angle diffraction patterns from the two states appeared to be identical, suggesting that a major conformational change of the myosin heads was not driven directly by ATP hydrolysis on the protein. This was later interpreted to indicate that significant changes of cross-bridge angle or conformation do not occur directly as a result of nucleotide interaction with the myosin active site, at least not in the absence of actin, but only as a result of interaction with thin filaments (Mannherz and Goody, 1976). This is therefore contrary to the idea of the recovery stroke proposed by Lymn & Taylor (1971) shown in the scheme of Fig. 1c. Whether this interpretation is correct or not is still an unanswered question and is the subject of numerous structural and spectroscopic studies (see Goody and Holmes, 1983, for a review). Recent electron microscope evidence suggests that there may indeed be large structural changes associated with the interaction of ATP with isolated thick filaments from insect flight muscle in the absence of actin, although it is not yet clear whether, and in what manner, these are related to the contractile cycle (Clarke et al., 1986).

The second result emerging from this early work concerned the interaction of another ATP analogue, AMPPNP (Yount et al., 1971), with insect flight muscle fibres. In this analogue, the bridging oxygen between the beta and gamma phosphate groups is replaced by an NH group. This modification results in complete resistance to hydrolysis by myosin or actomyosin ATPase (Yount at al., 1971 b). Addition of this analogue to muscle fibres in the rigor state, i.e. in the absence of ATP, led to an intermediate state which had structural characteristics reminiscent of both rigor and relaxed muscles and this structural change was accompanied by a loss of tension but not of stiffness (Barrington Leigh et al., 1973; Beinbrech et al., 1976; Marston et al., 1976). The x-ray diffraction changes were characterised using the DESY bench and photographic film, and later using X11 together with a televisiontype detector (Goody et al., 1975; Tregear et al., 1979), and an x-ray titration showed that these took place in the same concentration range as the mechanical change. These effects have been confirmed more recently (Poole, Maéda & Goody, unpublished) using a new 2-D multiwire detector (described by Maéda et al., 1986) and are illustrated in Fig. 4. The computerised image of Fig. 4a shows an example of a 1 minute exposure of the rigor state taken on the X33 bench, and the integrated intensity profiles of 4b, taken along the equator, the 38.5 nm 1st actin layer line and the 14.5 nm myosin line clearly illustrate the differences between the relaxed, AMPPNP and rigor states. These effects were originally interpreted to indicate that in the presence of the analogue,

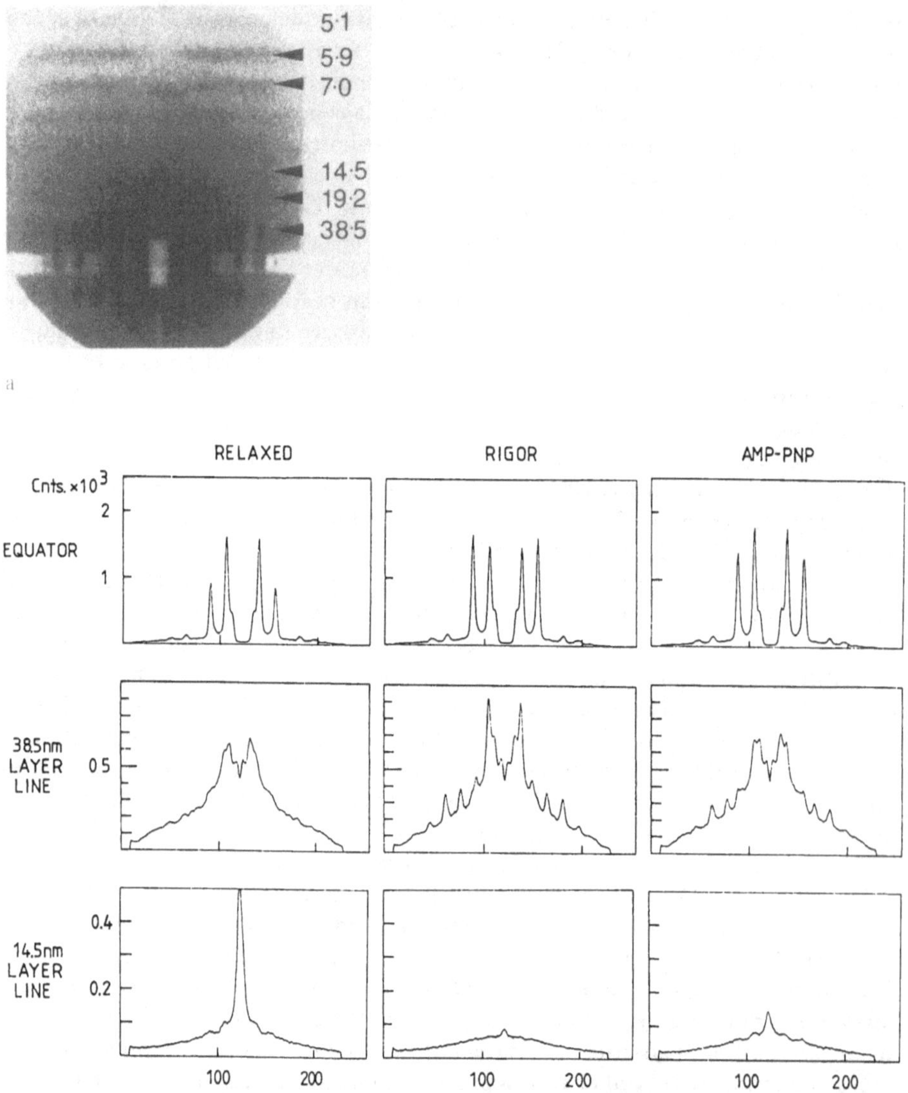

a

b

Fig. 4a and b. 2-D intensity data from a multiwire detector (detector details in Maéda et al., 1986). **a** shows a log intensity display of a 1 minute exposure of a 0.3 × 0.8 mm rigor fibre bundle, held horizontally in the DORIS, X33 camera (3.6 GeV, 50 mA) Reflections are broader in the meridional direction due to the horizontally elongated beam profile at X33 (see Table 1). **b** shows the lower angle data from relaxed (10 mM MgATP), rigor (no ATP) and AMPPNP (2 mM MgAMPPNP plus an ATP removal system containing hexokinase and glucose) patterns. The profiles represent intensities integrated over the whole width of the equatorial, 38.5 nm actin, and the 14.5 nm myosin layer lines. In AMPPNP the equatorial ratio is between relaxed and rigor values, the 38.5 nm reflections are all somewhat weaker but the 1,0 is particularly reduced, and the 14.5 nm intensity is increased. Poole, Maéda & Goody (unpublished)

myosin heads changed their orientation from ca. 45° in rigor to ca. 90° on binding AMPPNP without detaching from thin filaments, but it soon became clear from theoretical considerations that this simple explanation was unlikely (in particular, it is difficult to imagine how the incommensurate geometries of the thick and thin filaments would allow such a change; Barrington-Leigh et. al., 1977). Help was sought from electron microscopy, since a direct interpretation of the diffraction pattern was not possible. First results were interpreted to support the simple notion of a change of angle of bound cross-bridges (Marston et al., 1976), but closer inspection of the electron micrographs showed that the structure in the presumed AMPPNP state was not distinguishable from that in the presence of ATP under relaxing conditions.

This result drew attention to the potential artefacts in classical electron microscopy on muscle fibres, and resulted in a systematic study, using the DESY and X11 benches, of the structural consequences of fixation, staining, dehydrating and embedding in plastic which are necessary processes leading to the preparation of ultra-thin sections for electron microscopy (M. K. Reedy et al., 1983; M. C. Reedy et al., 1983). Using the X11 bench to rapidly monitor the structural state of the muscle fibres, it was demonstrated that artefacts occurred at the first stage of fixation, namely on fixation with glutaraldehyde. These artefacts were least serious for the rigor state and greatest in the relaxed state. In the case of AMPPNP, very slight variations in protocol determined whether the AMPPNP state was essentially preserved or whether a transition to the fixed (artefactual) version of the relaxed state occurred. This result probably explains the appearance of earlier electron micrographs from the AMPPNP state. Monitoring sample preparation by x-ray diffraction allowed the best preserved specimens to be selected for electron microscopy, and these have led to modification of the interpretations of the effect of AMPPNP on the structure. Briefly, although it is likely that a certain degree of cross-bridge dissociation does take place on addition of the analogue, AMPPNP does influence the structure of bound cross-bridges. In a recent contribution (Reedy et al., 1987), it is suggested that the combined electron microscope and x-ray evidence can best be explained on the basis of a two domain model of the myosin cross-bridge in which the two domains respond differently to interaction of the heads with thin filaments and nucleotides. On the assumption that the structural changes inferred are related to those occurring in the power stroke of muscular contraction, it is possible that the relevant structural change in the latter process is a change of relative disposition of these two domains rather than a change of angle of the cross-bridge without a large change in its internal structure.

Interestingly, Clarke et al. (1980) discovered, in a low temperature study of insect muscle, that similar mechanical and structural changes occur on addition of the antifreeze solvent ethylene glycol. They have since used the X13 x-ray bench to characterize the structural changes occurring on addition of this solvent at temperatures above zero and have compared them directly with those induced by AMPPNP (Clarke et al., 1984). They conclude that the structural and mechanical effects of these two agents are indistinguishable from each other, and suggest that they are particular cases of a general process of "equilibrium relaxation", although the mechanism of action of ethylene-glycol on the cross-bridge structure is not understood.

Using film and delay-line detectors on the DESY and X11 benches, a study of the binding of exogeneous myosin heads (S1 fragments) to insect flight muscle in the rigor state was carried out (Goody et al., 1985). Empty myosin-binding sites on thin filaments are occupied by the added S1 (isolated from rabbit muscle) and lead to characteristic changes in the the diffraction pattern. Since this happens over a period of several hours, and film exposures showing the whole diffraction pattern including the weaker diffuse actin-based outer layer lines took only a few minutes on X11, it was simple to follow the time dependence of this reaction and to determine when it was complete. Processing for electron microscopy was monitored by x-ray diffraction, and samples which showed the smallest changes relative to the unfixed material were used for preparing ultra-thin sections. From the combined x-ray and e.m. results, together with results from quantitative gel-electrophoresis and quantitative interference microscopy, the number of actin monomers still available for interaction with myosin heads in the rigor state were determined. Two principle interpretations could be made based on this evidence. Firstly, the number of myosin molecules per 14.5 nm repeat of the thick filament in this muscle is 4, in agreement with Wray, (1979) and Reedy et al., (1981), but in disagreement with much earlier evidence suggesting 6. Secondly, in the rigor state, 75–85% of the myosin heads are attached to thin filaments, in contrast with an earlier suggestion that only 50% are attached (Offer and Elliot, 1978), but in agreement with more recent evidence (Lovell et al., 1981). The total occupancy of actin monomers by myosin heads appeared to be very high, so that the parts of the diffraction pattern adhering to the thin filament symmetry could be considered to arise from a highly regular array of "decorated" actin. These diffraction data were used in modelling the shape of rigor cross-bridges (Holmes et al., 1980) and were also combined with phase data from Fourier analysis of e.m. images of decorated actin to refine a computer model of this structure (Holmes et al., 1982). The diffraction data were also used to guide the selection of the best preserved e.m. images of decorated actin which were then used for a 3-D reconstruction (Amos et al., 1982). In this work, it was suggested that a myosin head interacts with two actin monomers (while still retaining a 1 : 1 stoichiometry), but this point has not been proved definitively.

9 Dynamic Experiments

9.1 Stretch Activation Experiments

Some of the evidence for the attachment of cross bridges in the rigor state has been presented in the previous section, but, in all the muscles studied it has proved much more difficult to demonstrate a clear enhancement of intensity associated with the actin layer lines, signifying myosin binding, in the actively contracting state. Some enhancement of the 5.9 nm layer line has been measured in activated frog fibres by a number of groups (Matsubara et al., 1984; Wakabayashi et al., 1985; Kress et al., 1986; Maéda et al., 1987), but it is still not clear how much of this change is due to the thin filament activation process itself. Recently, however, Maéda et al. (1986), using a 2-dimensional multiwire detector at the Hamburg synchrotron, have succeeded in demonstrating a clear increase in the intensity of the first actin layer line on activation of king crab muscle.

Given the above difficulties, most of the information provided by x-ray diffraction about the structural events leading up to and associated with tension production has thus come from experiments in which the dynamics of the changes in the equatorial, meridional and myosin layer line intensities have been correlated with the dynamics of tension production. Much of this work has been done at synchrotron radiation sources by H. E. Huxley and his colleagues (see reviews by H. E. Huxley and Faruqi, 1983; Huxley, 1984; Kress et al. 1986) who have collected time resolved data during repeated electrical excitation of whole frog muscles. The stretch activatable fibrillar muscle from insects provides an alternative to this system which lends itself readily to repeated activation experiments on an x-ray beamline. Further, skinned preparations can be used which are less sensitive to radiation damage and allow the experimenter to control the constituents in the bathing medium.

Armitage, Miller, Rodger & Tregear (1972) (see also Armitage, Tregear & Miller, 1975) were the first to take advantage of these properties of fibrillar muscle in an early attempt to correlate changes in the x-ray scattering with stretch induced tension changes. They used a rotating anode source and phased the counting of x-rays arriving at proportional counters through focal plane apertures positioned over various reflections with a particular period in a forced oscillation of *Lethocerus* fibres, and accumulated data over several thousand cycles. Despite the elegance of these experiments they were necessarily handicapped by the limited x-ray intensity and detector capabilities. Such experiments have, of course, become much more feasible with the development of synchrotron radiation beamlines and rapid, multiwire position sensitive detectors, and recently an analogous series of experiments on insect muscle has begun at the X33, EMBL beamline in Hamburg (Rapp, Poole, Maéda, Güth & Goody, in preparation).

Fibre activation is associated with a large increase in ATPase activity (Rüegg & Stumpf, 1966) and so the size of the skinned fibre bundles used in such experiments is limited by the rate of diffusion of ATP into the fibres. Thus fewer than ten fibres were used, mounted between a tension transducer and a vibrator in the horizontal direction to maximize the intensity of the horizontally elongated X33 beam on the specimen (See Table 1 in the Appendix). The stretch activation process requires the presence of calcium ions, which alone have little or no effect on the relaxed tension (at least in recently extracted fibres), and figure 5 shows the changes we have measured in the equatorial and meridional parts of the pattern when fibres bathed in a calcium buffer were subjected to a series of rapid stretch-release cycles. Intensity data were collected in a series of time frames along with physiological parameters such as fibre tension and length and calibration parameters such as incident intensity using the CATY data accquisition system (Golding, 1982), provided by the EMBL outstation, which runs on an LSI 11/2 auxilliary crate controller linked to a PDP 11/45 computer via a CAMAC serial highway. The data were then dumped to the EMBL VAX-750 computer and analysed with the help of a package of programs especially designed to locate and quantify the intensity associated with reflection peaks (Koch & Bendall, 1981).

Parts a & c of Fig. 5 show 3-D plots of the changes in the 14.5 nm layer line and equatorial reflections measured during an activation cycle at 100 ms time resolution, and are to be compared with the tension transient shown in Fig. 5b. Clearly active tension production is associated with a large depression of the intensity of the 14.5 nm

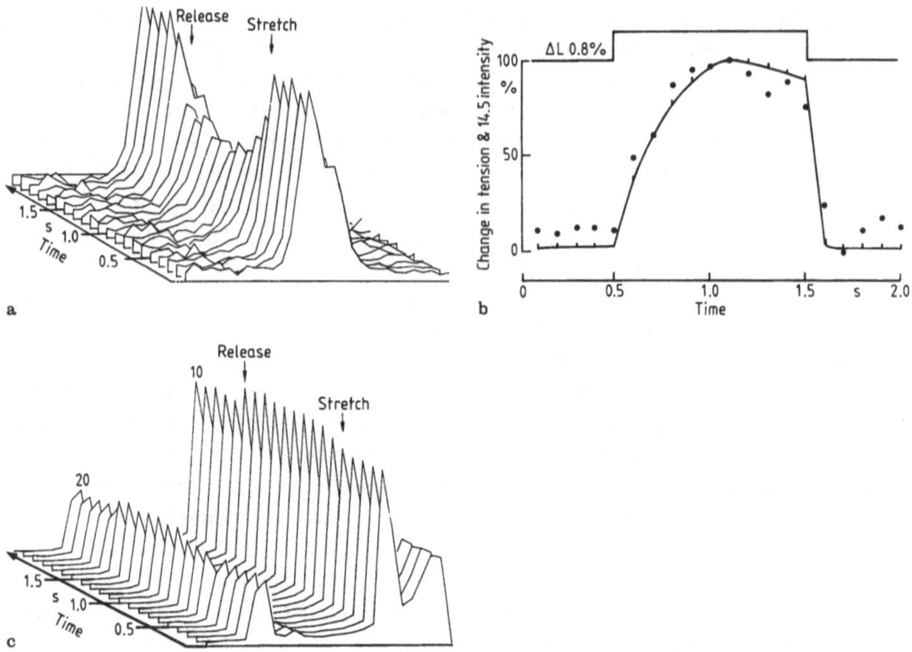

Fig. 5a–c. Structural changes on stretch activation. Conditions were, 10 °C, 10 mM MgATP plus an ATP backup system of 10 mM creatine phosphate and 1 mg/ml creatine phosphokinase, and 10^{-4} M calcium ions. **a** shows a 3-D plot of the changes in the 14.5 nm reflection measured using a linear multiwire detector placed along the 14.5 nm layer line. Each time frame is 100 ms and represents data accumulated over 40 stretch-release cycles, which included an intervening 4 second rest phase. **b** shows the time course of the tension rise and integrated intensity change (measured after background subtraction) of the experiment in **a**. **c** shows a 3-D plot of the changes in one half of the equatorial pettern measured during a similar cycle of stretch and release

reflection, but with very little change in the equatorial, 1.0 and 2.0 intensities. Figure 5b illustrates the close relationship between the time course of the change in the integrated intensity of the 14.5 nm peak and that of the tension rise. This was confirmed by the results of experiments done at higher temperature which showed a similar acceleration of both rates (Rapp et al., in preparation). In most experiments the width of the 14.5 reflection across the meridian remained unchanged and so disordering of the thick filament lattice does not appear to complicate the intensity change measured, unlike the situation found on activation of frog muscle (Huxley et al., 1982). Thus, in this preparation, it appears that cross-bridge binding, or at least tension production by cross-bridges, causes a loss of the 14.5 nm axial ordering of the myosin heads. This presumably indicates that binding to actin requires at least some part of the myosin head to align itself with the actin geometry to the detriment of the longitudinal organisation of bridges projecting from the thick filaments in relaxed muscle.

A knowledge of the relationship between the loss of 14.5 nm meridional intensity and the actual number of attaching heads would be very useful indeed as a measure of head attachment in other situations, e.g. in the ATP jump experiments described in the next section. Since the high frequency stiffness, thought to be a measure of the number of bridges attached (Huxley & Simmons, 1971), is known to increase linearly with the delayed tension production following a stretch (White et al., 1979), then a similar time course of meridional intensity change and tension change in the above experiment would indicate that the relationship is linear within this region of change. Looking at the data of Fig. 5b it appears that the change in the intensity of the 14.5 nm reflection is steeper than the tension rise suggesting that the relationship between head attachment and 14.5 nm meridional intensity is actually non-linear. The above experiment must be done at higher time resolution to characterize this relationship fully, but it would be better to correlate tension and stiffness measurements with the meridional intensity when fibres are activated to different degrees by varying amplitudes of stretch (Jewell & Rüegg, 1966; Güth et al., 1981).

The equatorial reflections show a small, $<1\%$, outward shift of their peak intensities on stretch activation, shown in Fig. 5c, indicating that a $<1\%$ reduction in lattice spacing occurs. However, the calculated change in the ratio of the 2,0 to 1,0 intensities is very small, actually $<10\%$ of the change measured when fibres go into rigor. Clearly there is a gross molecular reorganisation of mass within the lattice as 70–80% of the myosin heads bind to actin in the rigor state (Goody et al., 1985; Lovell et al., 1981), but care must be taken when interpreting the above equatorial intensity ratios in terms of the relative numbers of cross-bridges bound in these two situations. There are arguments suggesting that cross-bridge binding need not necessarily involve large radial movements of mass from the thick filaments (1,0 planes) towards the thin filaments (2,0 planes), but that changes in the angle and slew of bound heads may be more important in determining the average distribution of mass between the filaments and hence the equatorial intensity ratio (Lymn, 1978; Xu et al., 1987). The large fall in the 2,0/1,0 equatorial ratio on addition of AMPPNP to insect fibres in rigor, as seen in Fig. 4b of the previous section, illustrates this point since the high frequency stiffness measurements indicate that a similar number of bridges are bound in both equilibrium states (Marston et al., 1976). The surprisingly small change in the equatorial intensities on activation of this muscle may therefore result either from the attachment of an equally small proportion of the number attaching in rigor, or from a larger number whose average disposition of mass between the thick and thin filaments is not very different from that found in the relaxed state. The answer is not yet clear although the aforementioned stiffness measurements made during activation show that the value in maximally activated fibres reaches 50% (Poole, 1984), or more (White & Mulloy, unpublished) of the value in rigor, suggesting that a relatively large number of bridges are attached. This assumes, of course, that the stiffness of an individual cross-bridge remains the same in the two situations.

9.2 ATP-Jump Experiments

A combination of the novel flash photolysis technique enabling the rapid release of nucleotides such as ATP from inert, photolabile precursors (Kaplan et al., 1978; McCray et al., 1980; Gurney & Lester, 1987) with the high x-ray intensities available from synchrotron sources has introduced the possibility of studying the kinetics of structural events associated with the actomyosin ATPase reactions in muscle fibres. The precursor or "caged" nucleotides can readily diffuse into skinned muscle fibres, where, in the case of caged-ATP, a pulse of ultraviolet (u.v.) light will photolyse

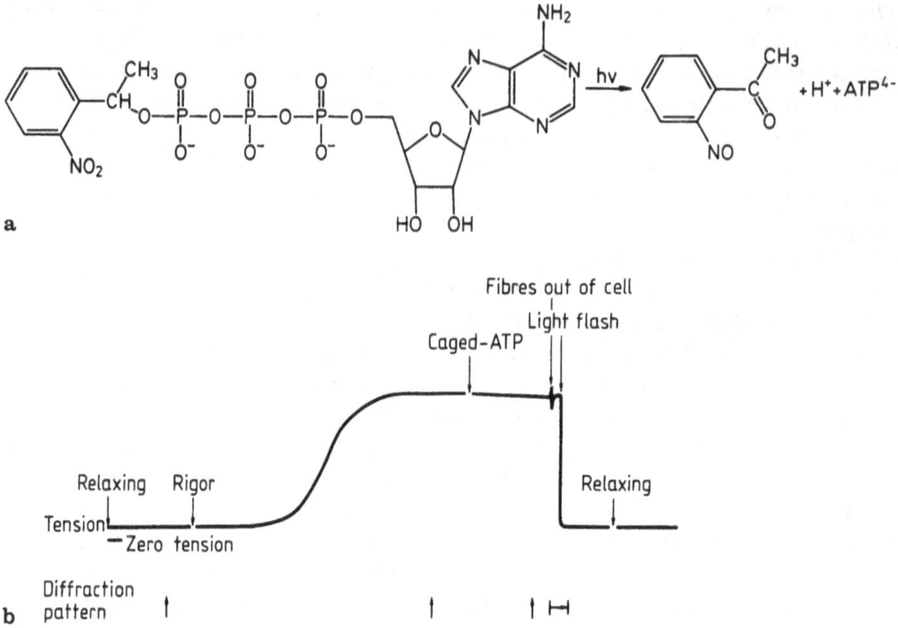

Fig. 6a and b. The use of caged-nucleotides. **a** Caged-ATP, P^3-1-(2-nitro)phenylethyladenosine 5'-triphosphate, a photolabile precursor of ATP. It can be cleaved by u.v. light from a laser or xenon flash lamp. The rate of ATP release is limited by a dark reaction which is pH sensitive (McCray et al., 1980). At pH 6.8, 25 °C it is ca. 350 s^{-1}. The deleterious oxidative effects of the nitrosoacetophenone leaving group can be countered by the addition of dithiothreitol reducing agent (DTT) to the solution bathing the fibres. **b** shows the experimental protocol of an ATP jump experiment. Diffraction patterns are taken at the points indicated by the arrows below the figure. The development of the rigor state is rate limiting in this sequence

the molecule releasing ATP at a rate of a few hundred per second (at pH ca. 6.8 at 25 °C) together with a nitrosoacetophenone leaving group as shown in Fig. 6a. Consequently, the diffusion problem preventing the rapid mixing of substances within whole fibres has now been surmounted and a number of studies using caged-ATP have already reported on the kinetics of the complex tension relaxation (Goldman et al., 1984; Hibberd & Trentham, 1986) and the biochemical changes (Ferenczi et al., 1986) which occur following the rapid ATP release within rigor fibres.

The first time resolved diffraction studies using this compound were performed on insect muscle at the X33 beamline at the DORIS storage ring, Hamburg, and used a xenon flash lamp (Rapp and Güth, 1988) as a source of u.v. light for the photolysis (Goody et al., 1985; Rapp et al., 1986). Bundles of around 40 fibres were used, the size being limited by factors such as the diffusion times required for the removal of ATP and the entry of caged-ATP, and the extent of u.v. light penetration across the bundle. It was not possible, as in the stretch activation experiments, to accumulate data over many ATP-jump cycles due to the diffusion delays, so the x-ray absorption was reduced as far as possible during the periods of dynamic data collection by temporarily suspending the fibres in air. In this way adequate equatorial and meridional signals could be collected from single cycles with 1–5 ms time resolution using the 1-D multiwire detector when the x-ray source conditions were optimal, i.e. with a high current and high beam stability in the frequency range of the measurements. The details of the beamline optics on X33 are described in the Appendix. The fibres were mounted horizontally in the beam and held isometric throughout the experiment.

The diagramatic tension trace of Fig. 6b illustrates the experimental protocol; 1 sec control exposures were taken in the relaxed state and after rigor tension generation, and dynamic data were collected in a series of up to 256 time frames of variable length starting just before the light flash. Usually the first 150 or so frames were between 1–5 ms long to monitor structural changes taking place on a fast time scale, and were followed by a number of longer frames during the remainder of the tension relaxation. Fig. 7a illustrates the quality of the equatorial data collected at 1 ms time resolution, and demonstrates the dramatic effect of ATP release on the rigor pattern. The integrated intensity ratio of the 2,0 : 1,0 reflections is plotted alongside the tension relaxation in Fig. 7b and shows that a gross structural change is occurring on a much faster time scale than the tension change. If this movement of rigor cross-bridges on ATP binding reflects cross-bridge detachment, then a comparison of these data with the simulated detachment curves of Fig. 7c indicates that the apparent second rate constant of detachment, K_d in the scheme shown, is ca. $10^6\ M^{-1}\ s^{-1}$ which is very similar to the rate of actomyosin dissociation by ATP in solution (Lymn & Taylor, 1971; White et al., 1987). There does, however, appear to be some inconsistency here since the structural change evidenced by the equatorials is complete within a few milliseconds whereas the tension is still relatively high at this time and takes a few hundred milliseconds to relax fully. In order to explain a similarly slow tension relaxation of rabbit psoas fibres, Goldman et al. (1984) have proposed that some cross-bridges may cooperatively reattach after an initial rapid detachment on ATP binding, and that these actively cycling bridges generate tension which presumably decays at a rate determined by a subsequent, rate limiting step on the attached pathway. Our equatorial results are consistent with such a model since we measure a very rapid change which could be cross-bridge detachment, and we know from the activation experiments described in the previous section that the binding of cycling bridges in that situation had very little measurable effect on the 1,0 and 2,0 intensities. Thus we may not necessarily expect to measure an equatorial transient corresponding to the behaviour of the proposed population of reattached cycling bridges in the caged-ATP experiment.

The strong 14.5 nm periodicity of the axial organisation of the cross-bridges in relaxed muscle is, however, very clearly sensitive to the binding of active, tension producing bridges, as seen in the Fig. 5b, so we predicted that in the caged-ATP experiments this reflection should show a slower phase of intensity increase corresponding to the discharge of the supposed reattached population. Figure 8a shows a 3-D plot of the data collected at 5 ms time resolution, and the time course of the integrated intensity change, seen in Fig. 8b, shows that after an initial jump, the intensity of the 14.5 nm meridional reflection does, indeed, show a slower increase which has a similar time course to the tension fall. The measurement of this intensity transient with the detector placed along the 14.5 nm layer line confirmed that the changes are not associated with any significant width changes and so are likely to reflect changes in the longitudinal ordering of bridges rather than disturbances in

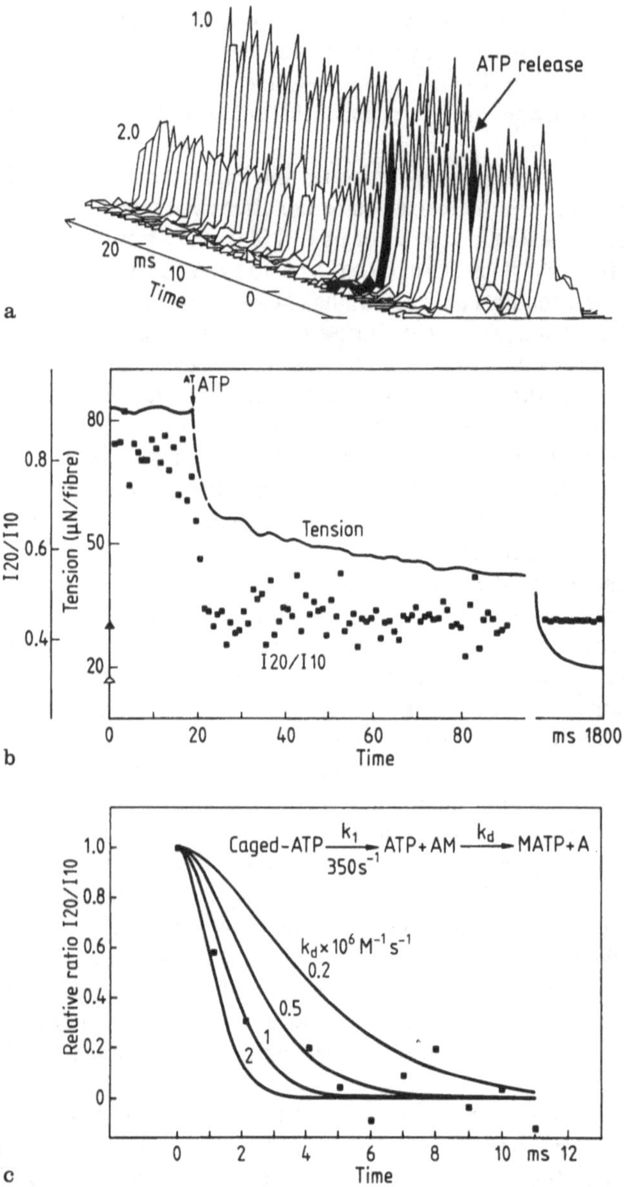

Fig. 7a–c. Changes in the equatorial pattern following the rapid release of ca. 2 mM ATP. **a** shows a 3D plot of the changes in one half of the equatorial pattern. Each time frame is 1 ms and the darkened frame indicates the time of the lamp flash. The fibres were initially in a rigor solution (10 mM $MgCl_2$, 5 mM EGTA, 45 mM KCl, 20 mM MOPS, 5 mM NaN_3, pH 6.8) containing 10 mM caged-ATP + 10 mM DDT. **b** shows the time course of the changes in tension and ratio of the integrated intensities of the 2,0 and 1,0 reflections (background does not change during the transient so has not been subtracted). The data were collected in 1 ms time frames and later in longer ones. The symbols on the ordinate indicate the values of relaxed tension (open triangle) and relaxed equatorial intensity ratio (solid triangle). The half time of the final tension relaxation is ca. 50 ms. The lamp flash caused an artefact on the tension record which has been replaced by the dashed trace, obtained from subsequent

the myosin filament lattice. The amplitude of the slower phase varied from experiment to experiment, averaging between 20% and 50% of the complete rigor-relaxed change. A comparison with the 40% depression seen on stretch activation may suggest that the reattaching population could be as large as that measured in the fully activated fibres.

Similar experiments have been performed more recently using rabbit psoas muscle fibres (Poole et al., 1987a & b), where the equatorial intensities are known to be sensitive to the number of actively cycling bridges bound (Brenner & Yu, 1983).

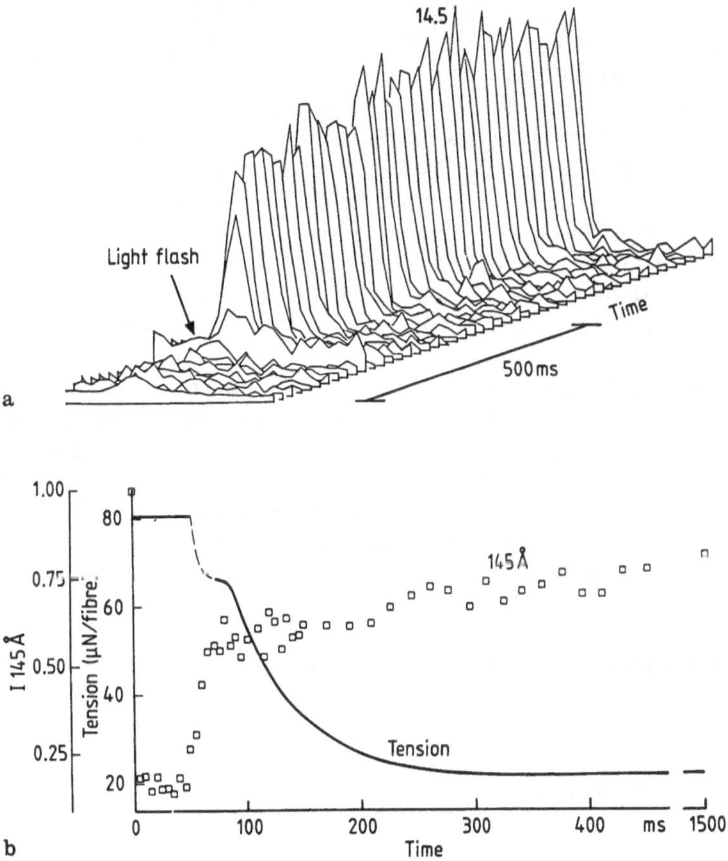

Fig. 8a and b. Changes in the meridional diffraction pattern measured at 5 ms time resolution after rapid ATP release. **a** 3-D plot of one of the two 14.5 nm peaks on the meridian showing the rapid changes ocurring on release of ca. 1 mM ATP. **b** The two phase time course of the integrated intensity change. The final point on thc right hand side indicates the value measured in relaxed fibres. From Rapp ct al. (1986)

◀ experiments performed under similar conditions. **c** shows simulations of the time course of ATP induced detachment of rigor bridges for different values of k_d in the scheme shown. The data are from **b**. The intensity ratio has been normalised to a value of 1 in rigor and 0 in the relaxed states. With a value of 350 s^{-1} for k_1 the best fit gives a value of ca. 10^6 M^{-1} s^{-1}. From Poole et al. (1987)

In these experiments the equatorial intensities show a two-phase change following the rapid release of ATP, unlike the insect equatorials but similar to the time course of the change in intensity of the insect 14.5 nm reflection. Thus the equatorials of this muscle have provided us with an opportunity of monitoring the behaviour of the re-attached population of bridges under a variety of conditions e.g. in the presence and absence of phosphate ions (Poole et al., 1987), at high time resolution. Thus, it is important to remember that diffraction patterns from different muscle types have many common features but also many differences, due to variations in lattice type, thick filament structure and regulatory protein content and arrangement (Squire, 1981; Wray & Holmes, 1981), and that they may show different intensity changes when cross-bridges bind. It is therefore important to choose the muscle specimen appropriate to the particular intensity measurement being made.

These experiments represent only the beginning of a whole range of kinetic experiments now possible using the flash photolysis technique. A new method of synthesising caged nucleotides, developed recently by Walker, Ried, McCray & Trentham (personal communication), has made it possible to prepare caged nucleotide analogues such as caged-AMPPNP and caged-ATP(γ-S), and kinetic structural measurements of events occurring on binding of these analogues in insect muscle have already started (Poole, Rapp, Maéda & Goody, unpublished). Such kinetic data are likely to give us a new insight into the nature of the structural changes taking place in the cross-bridges when these nucleotides bind.

The very recent development of a caged-calcium molecule (Tsien, 1986; Kaplan & Ellis-Davies, 1988) is also of great interest in the study of the kinetics of cross-bridge binding following activation. Its application in the study of fibre types activated purely by calcium, such as vertebrate striated muscle, is clear, and we have recently measured the rate of equatorial intensity changes in rabbit psoas fibres on release of Ca^{2+} from its complex with the EDTA derivative DM-nitrophen ($t_{1/2}$ = ca. 20 msec at 25 °C). However, it may also prove interesting to use such a compound in the investigation of the mechanism of stretch activation in insect flight muscle. This should be possible by measuring the kinetics of the mechanical and structural changes occurring when calcium is released into fibres held at different degrees of stretch.

10 Conclusion and Prospects

The development of synchrotron radiation as a source for x-ray diffraction and the investigation of the structure and contractile mechanism of insect flight muscle have had a symbiotic relationship since the early 1970's. While it is perhaps true that, until a few years ago, synchrotron radiation research had benefitted more from this than insect flight muscle research, recent time-resolved work has begun to fulfill the promise of those first diffraction experiments carried out by Rosenbaum et al. (1971) in 1970.

Further progress in time resolved experiments will need increased x-ray intensity and smaller beam sizes, particularly for measurement of the weaker reflections from insect flight muscle. Fortunately, both of these improvements are expected over the next few years from new storage ring sources, from wigglers and undulators and from better optical arrangements. In parallel, there is an acute need for more suitable detectors, since those available are not able to take full advantage of present source intensities. Again, current developments in 2D detector technology also suggest that these will become available in the near future.

11 Appendix

11.1 X-Ray Optics for Small Specimens

Here we consider some of the principle features of x-ray optics for beam lines on synchrotron radiation sources, with particular reference to the special requirements of small specimens. The most important factors involved are the size and position of the virtual source, the distance between the virtual source and the focussing elements relative to that between the focussing elements and the focus, and the presence and performance of the focussing systems. These points are considered briefly below.

1) The virtual source is the point from which the radiation appears to originate, and is generally not at the actual source (tangent point) on the electron or positron orbit. The virtual source can be upstream or downstream of the actual source. The relative position of the virtual source differs greatly around the orbit and is also a function of the operating mode of the ring. The position and size of the virtual source are critically important in determining the size of the focus, as can be seen from the following expression for the size in a particular direction

$$s_{x,z} = S_{x,z}v/u$$

where $s_{x,z}$ = the size of the focus in the horizontal (x) or vertical (z) direction, $S_{x,z}$ = size of the virtual source, u = source-mirror or monochromator distance, v = mirror- or monochromator-focus distance. The influence of these parameters on the predicted size of the focus, assuming ideal focussing optics, is seen in the comparison of X13 and X33 given in Table 1.

As shown in the table, the smallest focal size potentially achievable with X13 is considerably smaller than with X33 in the vertical direction, although not in the horizontal direction. Nevertheless, the smaller vertical size means that a much larger fraction of the total flux could be concentrated on a thin specimen using X13 with focussing mirrors than with X33.

2) If flat mirrors are used, as was the case with X11 and X13 and is now the case with X33, the size of the pseudo-focus is given by adding a further term to the expression given above which is essentially a magnified mirror aperture term (for a single mirror). The complete expression for this situation is:

$$s_z = S_z v/u + a(u + v)/u$$

Where a is the aperture made by a single mirror. On the basis of this expression, it is of interest to calculate the effect of focussing on the size of the vertical focus on the currently used bench X33. Taking approximate values for s_z (3 mm), u (22 m) and v (7 m) from Table 1, and with a = 0.7 mm, the contributions of the first term

Table 1. Optical Properties of Beam Lines X13 and X33 under Different Operating Modes

	X13(ELUM)	X33(ELUM)	X13(SYN)	X33(SYN)
		vertical		
S_z (mm)	1.6	3.2	1.1	2.2
S_z' (mrad)	0.64	0.38	0.53	0.34
u_0 (m)	-3.2	1.1	-2.3	1.1
u'	23.5	23.5	23.5	23.5
$u = u' - u_0$	26.7	22.4	25.8	22.4
v (m)	7	7	7	7
$c = u/v$	3.8	3.2	3.7	3.2
$s_z = S_z/c$ (mm)*	0.4	1.0	0.3	0.7
		horizontal		
S_x (mm)	5.1	5.3	3.9	4.0
S_x' (mrad)	2.0	2.0	1.3	1.4
u_0 (m)	-0.1	3.8	-0.08	3.2
u' (m)	21	21.5	21	21.5
$u = u' - u_0$	21.1	12.7	21.1	18.3
v (m)	9	9	9	9
$c = u/v$	2.3	2.0	2.3	2.0
$s_x = S_x/c$ (mm)	2.2	2.7	1.7	2.0

S_z, S_x size of the virtual source (4 sigma)
S', S_x' divergence of the source (4 sigma)
u_0 actual to virtual source distance (+, downstream)
u' actual source to lens** distance
u virtual source to lens distance
v lens to focus distance
c demagnification factor
s focus size*
* a flat mirror would give a larger focus
** focussing mirror or monochromator
ELUM-mode 5.1 GeV, electrons and positrons
SYN-mode 3.8 GeV, electrons only

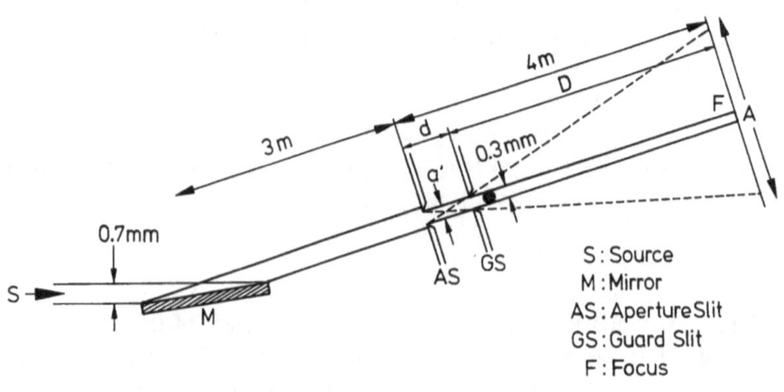

Fig. 9. Limitation of the beam in the vertical direction. See appendix

(0.95 mm) and the second term (0.92 mm) are almost identical. Thus, using focussing mirrors to eliminate the second term, the focal size in the direction would be halved. 3) The large size of the beam at the specimen in the vertical direction on X33 means that the vertical aperture must be limited to obtain an acceptable signal to noise ratio when using thin specimens. Since the beam is still divergent after reflection from the flat mirrors, beam size limitation must be done just in front of the specimen. This introduces the additional problem of scattering from the aperture slits, and this can only be overcome by the inclusion of a further set of guard slits immediately before the specimen (see Fig. 9). The distance between the aperture and the guard slits must be chosen so that the strong central scattering is smaller than the beam stop placed just in front of the detector. The height of this scattering can be calculated using the geometriccal construction shown in Fig. 9 to be given by

$$A = a'(1 + 2D/d)$$

Obviously, focussing optics would concentrate more of the beam onto a small specimen, and perhaps obviate the need for such complicated slit arrangements which are awkward and time-consuming to adjust.

12 References

Amos, L. A., Huxley, H. E., Holmes, K. C., Goody, R. S., Taylor, K. A.: Nature (London) *299*, 467–469 (1982)

Armitage, P. M., Miller, A., Rodger, C. D., Tregear, R. T.: Cold Spring Harbour Quant. Biol. *37*, 379–87 (1973)

Armitage, P. M., Tregear, R. T., Miller, A.: J. Mol. Biol. *92*, 39–53 (1975)

Bagshaw, C. R., Eccleston, J. F., Trentham, D. R., Yates, D., Goody, R. S.: Cold Spring Harbor Symp. Quant. Biol. *37*, 127–135 (1973)

Barrington-Leigh, J., Goody, R. S., Hofmann, W., Holmes, K. C., Mannherz, H. G., Rosenbaum, G., Tregear, R. T.: In Tregear, 1977, 137–146 (1977)

Barrington-Leigh, J., Holmes, K. C., Mannherz, H. G., Rosenbaum, G., Eckstein, F., Goody, R. S.: Cold Spring Harbour Symp. Quant. Biol. *37*, 443–448 (1973)

Barrington-Leigh, J., Rosenbaum, G.: Ann. Rev. Biophys. Bioeng. *5*, 239–270 (1976)

Beinbrech, G., Kuhn, H. J., Herzig, J. W., Rüegg, J. C.: Cytobiology *12*, 385–396 (1976)

Brenner, B., Yu, L. C.: Biophys. J. *41*, 257a (1983)

Clarke, M. L., Hofmann, W., Wray, J. S.: J. Mol. Biol. *191*, 581–585 (1986)

Clarke, M. L., Rodger, C. D., Tregear, R. T., Bordas, J., Koch, M.: J. Musc. Res. & Cell Motil *1*, 195–196 (1980)

Clarke, M. L., Rodger, C. D., Tregear, R. T.: J. Musc. Res. & Cell Motil. *5*, 81–96 (1984)

Eisenberg, E.: Lectures on Mathematics in the Life Sciences *16*, 19–59 (1986)

Ferenczi, M. A., Homsher, E., Trentham, D. R.: J. Physiol., London *352*, 575–599 (1984)

Golding, F. R.: Nucl. Inst. Meth. *201*, 231–235 (1982)

Goldman, Y. E., Hibberd, M. G., Trentham, D. R.: J. Physiol., London *354*, 577–604 (1984)

Goody, R. S., Barrington-Leigh, J., Mannherz, H. G., Tregear, R. T., Rosenbaum, G.: Nature (London) *262*, 613–615 (1976)

Goody, R. S., Eckstein, F.: J. Amer. Chem. Soc. *93*, 6252 (1971)

Goody, R. S., Holmes, K. C.: Biochim. Biophys. Acta *726*, 13–39 (1983)

Goody, R. S., Holmes, K. C., Mannherz, H. G., Barrington-Leigh, J., Rosenbaum, G.: Biophys. J. *15*, 687–705 (1975)

Goody, R. S., Reedy, M. K., Hofmann, W., Holmes, K. C., Reedy, M. K.: Biophys. J. *47*, 151–169 (1985)

Goody, R. S., Güth, K., Maéda, Y., Poole, K. J. V., Rapp, G.: J. Physiol. London *364*, 75P (1985)
Gurney, A. M., Lester, H. A.: Physiol. Rev. *67*, 583–617 (1987)
Güth, K., Kuhn, H. J., Tsuchiya, T., Rüegg, J. C.: Biophys. Struct. Mech. *7*, 139–169 (1981)
Hendrix, J., Fürst, H., Hartfield, B., Dainton, D.: Nuclear Instr. Meths. *201*, 139–144 (1982)
Hendrix, J., Koch, M. H. J., Bordas, J.: J. Appl. Crystallogr. *12*, 467–472 (1979)
Hibberd, M. G., Trentham, D. R.: Ann. Rev. Biophys. Biophys. Chem. *15*, 119–161 (1986)
Holmes, K. C., Blow, D.: The Use of X-ray Diffraction in the study of Protein and Nucleic Acid Structure, Wiley, New York (1966)
Holmes, K. C., Goody, R. S., Amos, L. A.: Ultramicroscopy *9*, 37–44 (1982)
Holmes, K. C., Tregear, R. T., Barrington-Leigh, J.: Proc. R. Soc. Lond. B. Biol. Sci. *207*, 13–33 (1980)
Huxley, A. F.: Prog. Biophys. Chem. *7*, 255–318 (1957)
Huxley, A. F., Simmons, R. M.: Nature (London) *233*, 533–538 (1971)
Huxley, H. E.: Science *164*, 1356–1366 (1969)
Huxley, H. E.: In: Contractile Mechanisms in Muscle (eds. Pollack, G. H., Sugi, H.), Plenum Press (1984)
Huxley, H. E., Brown, H.: J. Mol. Biol. *30*, 383–434 (1967)
Huxley, H. E., Faruqui, A. R.: Ann. Rev. Biophys. Bioeng. *12*, 381–417 (1983)
Huxley, H. E., Faruqui, A. R., Kress, M., Bordas, J., Koch, H. J.: J. Mol. Biol. *158*, 637–684 (1982)
Huxley, H. E., Kress, M.: J. Muscle Res. & Cell Motil. *6*, 153–161 (1985)
Jewell, B. R., Rüegg, J. C.: Proc. Roy. Soc. London Ser. B *164*, 429–459 (1966)
Kaplan, J. A., Forbush, B., Hoffman, J. F.: Biochem. *17*, 1929–1935 (1978)
Koch, M. H. J., Bendall, P. J.: Proc. Digital. Equip. Comp. Users Soc. U.K. (University of Warwick), 13–16 (1981)
Kodama, T.: Physiol. Rev. *65*, 467–551 (1985)
Kress, M., Huxley, H. E., Faruqui, A. R., Hendrix, J.: J. Mol. Biol. *188*, 325–342 (1986)
Lehman, W.: In Tregear *1977*, 277–285 (1977)
Lemonnier, M., Fourme, R., Rousseaux, F., Kahn, R.: Nucl. Instrum. Methods *152*, 173–177 (1978)
Lovell, S. J., Knight, P. J., Harrington, W. F.: Nature (London) *293*, 664–666 (1981)
Lymn, R. W.: Biophys. J. *21*, 93–98 (1978)
Lymn, R. W., Taylor, E. W.: Biochem. *10*, 4617–4624 (1971)
Maéda, Y., Boulin, C., Gabriel, A., Sumner, I., Koch, M. H.: Biophys. J. *50*, 1035–1042 (1986)
Maéda, Y., Popp, D., McLaughlin, S.: in: Molecular Mechanism of Muscle Contraction (eds. Sugi, H., Pollack, G. H.) in press 1987
Mannherz, H. G., Goody, R. S.: Ann. Rev. of Biochem. *45*, 427–465 (1976)
Marston, S. B., Rodger, C. D., Tregear, R. T.: J. Mol. Biol. *104*, 263–276 (1976)
Matsubara, I., Yagi, N., Miura, Y., Ozeki, M., Izumi, T.: Nature (London) *312*, 471–473 (1984)
McCray, J. A., Herbette, L., Kihara, T., Trentham, D. R.: Proc. Natl. Acad. Sci. U.S.A. *77*, 7237–7241 (1978)
Miller, A., Tregear, R. T.: J. Mol. Biol. *70*, 85–104 (1972)
Miyahara, J., Takahashi, K., Amemiya, Y., Kamiya, N., Satow, Y.: Nucl. Inst. Meth. *A 246*, 572–578 (1986)
Offer, G., Elliott, A.: Nature (London) *271*, 325–329 (1978)
Poole, K. J. V.: Cross-Bridge Kinetics in Insect Flight Muscle, D. Phil Thesis, University of York, U.K. (1984)
Poole, K. J. V., Rapp, G., Maéda, Y., Goody, R. S.: J. Musc. Res & Cell Motil. *8*, 62a (1987a)
Poole, K. J. V., Rapp, G., Maéda, Y., Goody, R. S.: in: Molecular Mechanism of Muscle Contraction (ed. Sugi, H., Pollack, G. H.) (in press 1987)
Poulsen, F., Lowy, J., Cooke, P. H., Bartels, E. M., Elliot, G. F., Hughes, R. A.: Biophys. J. *51*, 959–968 (1987)
Pringle, J. W. S.: In Tregear *1977*, 177–196 (1977)
Rapp, G., Güth, K.: Pflügers Arch. *411*, 200–203 (1988)
Rapp, G., Poole, K. J. V., Maéda, Y., Güth, K., Hendrix, J., Goody, R. S.: Biophys. J. *50*, 993–997 (1986)
Reedy, M. C., Reedy, M. K., Goody, R. S.: J. Mus. Res. & Cell. Motility *4*, 55–81 (1983)
Reedy, M. C., Reedy, M. K., Goody, R. S.: J. Muscle Res. & Cell. Motil. *8*, 473–503 (1987)
Reedy, M. K.: Am. Zoologist *7*, 465–481 (1967)

Reedy, M. K.: J. Mol. Biol. *31*, 153–176 (1968)

Reedy, M. K., Garrett, W. E.: In Tregear *1977*, 115–136 (1977)

Reedy, M. K., Goody, R. S., Hofmann, W., Rosenbaum, G.: J. Mus. Res. & Cell. Motility *4*, 25–53 (1983)

Reedy, M. K., Holmes, K. C., Tregear, R. T.: Nature *207*, 1276–1280 (1965)

Reedy, M. K., Leonard, K. R., Freeman, R., Arad, T.: J. Musc. Res. & Cell. Motil *2*, 45–64 (1981)

Rosenbaum, G.: Die Nutzung der Synchrotronstrahlung für die Röntgenstrukturanalyse in der Molekularbiologie, Ph. D. Thesis, University of Heidelberg (1979)

Rosenbaum, G., Harmsen, A.: SSRL Rep. No. 78/04, VIII, 36–37, Stanford Linear Accelerator Center, California (1978)

Rosenbaum, G., Holmes, K. C.: In: Synchrotron Radiation Research (eds. Winick, H., Doniach, S.), Plenum Publishing Corporation 533–564 (1980)

Rosenbaum, G., Holmes, K. C., Witz, J.: Nature (London) *230*, 129–131 (1971)

Rüegg, J. C., Stumpf, H.: Pflügers Arch. *305*, 34–46 (1969)

Squire, J.: In: "Insect Flight Muscle, Proc. Oxford Symp" ed. R. T. Tregear, North Holland, Amsterdam, pp. 91–112 (1977)

Squire, J.: The Structural Basis of Muscle Contraction, Plenum Press, New York (1981)

Taylor, E. W.: CRC Crit. Rev. Biochem. *6*, 103–164 (1979)

Tregear, R. T.: In: Skeletal Muscle, American Handbook Physiol. 487–506 (1983)

Tregear, R. T.: Ed. Insect Flight Muscle: Proceedings of the Oxford Symposium, Amsterdam: North Holland (1977)

Tregear, R. T.: In: Development and Specialisation of Skeletal Muscle (ed. Goldspink, D.), Cambridge University Press, Cambridge 107–122 (1981)

Tregear, R. T., Milch, J., Goody, R. S., Holmes, K. C., Rodger, C. D.: In: Cross-bridge Mechanism in Muscle Contraction (eds. Sugi, H., Pollack, G. H.), University Park Press, Baltimore 407–423 (1979)

Tsien, R. Y.: In: Optical Methods in Cell Physiology (eds. DeWeer, P., Salzberg, B.), New York: Wiley (Soc. Gen. Physiol. Ser. *40*, 327–346 (1986)

Wakabayashi, K., Tanaka, H., Amemiya, Y., Fujishima, A., Kobayashi, T., Toshiaki, T., Sugi, H., Mitsui, T.: Biophys. J. *47*, 847–850 (1985)

White, D. C. S.: J. Physiol. London *208*, 583–605 (1970)

White, D. C. S., Wilson, M. G. A., Thorson, J.: In: Cross-bridge Mechanism in Muscle Contraction (eds. Sugi, H., Pollack, G. H.) University Park Press, Baltimore (1979)

White, D. C. S., Thorson, J.: J. Physiol., London *343*, 59–84 (1983)

White, D. C. S., Zimmerman, R. W., Trentham, D. R.: J. Musc. Res. & Cell. Motil. *7*, 179–192 (1986)

Winkelman, D. A., Mekeel, H., Rayment, I.: J. Mol. Biol. *181*, 487–501 (1985)

Wray, J. S.: Nature (London) *277*, 37–40 (1979)

Wray, J. S., Holmes, K. C.: Ann. Rev. Physiol. *43*, 553–565 (1981)

Wray, J. S., Vibert, P. J., Cohen, C.: Nature *257*, 561–564 (1975)

Xu, S., Kress, M., Huxley, H. E.: J. Muscle Res. & Cell. Motil. *8*, 39–54 (1987)

Yount, R. G., Babcock, D., Ballantyne, W., Ojala, D.: Biochem. *10*, 2484–2489 (1971a)

Yount, R. G., Ojala, D., Babcock, D.: Biochem. *10*, 2490–2496 (1971b)

Protein Single Crystal Diffraction

Ian D. Glover[1,2], John R. Helliwell[2,3] and Miroslav Z. Papiz[2]

Table of Contents

[1] Department of Crystallography, Birkbeck College, Malet Street, London WC1E 7HX, England
[2] Science and Engineering Research Council, Daresbury Laboratory, Daresbury, Warrington, Cheshire WA4 4AD, England
[3] Department of Physics, University of York, Heslington, York YO1 5DD, England

1 Introduction

The x-ray crystallographic analysis of protein structures is a remarkably successful technique. Since the publication of the first protein structure, that of myoglobin in 1958, many other protein structures have been determined. The resulting structural details often approaching atomic level have led to great insights into enzyme catalysis, hormone function, the organisation of the immune system, the molecular architecture of virus particles and protein synthesis. Why then should such an apparently successful technique need synchrotron radiation?

Although great progress has been made the understanding of biological systems at the molecular level is still at a very early stage. A vast amount of work has yet to be carried out to establish a complete molecular anatomy of these systems. There are many millions of proteins of differing functions interacting with other biological macromolecules. The conformation of a molecule is to some extent dependent upon its environment, in terms of ion, substrate or inhibitor interactions, and a series of slightly differing structures may be feasible depending upon these interactions. Genetic variation between species and generations may result in subtle structural alterations. Finally the mushrooming of biological and protein engineering, especially the production of specific site mutations in enzymes for example, makes the characterisation of such small structural changes an important area.

The solution of a new protein structure is very labour intensive and the relative ease with which a structure is determined is highly dependent upon the state of available technology. Due to its characteristics of intensity and tuneability synchrotron radiation is a special landmark in the state of technology applied as a research tool to the solution of protein structures.

It is feasible with synchrotron radiation to rapidly collect high resolution data, essential for the precise definition and refinement of molecular models, even from very large structures such as viruses. Using SR it is possible in many cases to collect complete datasets on a single sample, avoiding sample to sample variations. The high intensity is exploited in the collection of statistically significant data as a function of time; this approach allows the imaging of dynamic and/or structural intermediates involved in protein function. The smooth wavelength continuum of synchrotron radiation allows novel methods of solving the phase problem to be used in protein crystallography. The traditional method of multiple isomorphous replacement may be supplanted by selecting wavelengths to optimise the magnitudes of anomalous scattering effects for use in phase determination. In the case of a metalloprotein, derivatives need not be prepared at all.

2 Fundamentals of the Protein Crystallographic Technique

Bragg's law predicts the angle of reflection of any diffracted ray from specific atomic planes whereby

$$n\lambda = 2d \sin \theta$$

where d is the interplanar spacing of that set of planes, λ, is the wavelength of the

X-rays and n is an integer. The closer the separation of the planes then the larger the value of θ, the diffraction angle for a given λ; this corresponds to the higher resolution data.

Bragg's law is a special case of the Laue equations which define the condition for diffraction (constructive interference) to occur:

$$\mathbf{a} \cdot \mathbf{S} = h$$

$$\mathbf{b} \cdot \mathbf{S} = k$$

$$\mathbf{c} \cdot \mathbf{S} = l$$

where h, k, l are the Miller indices defining a unique plane of reflection (or diffraction spot), **a**, **b**, and **c** are the lattice vectors and **S** is the vector path difference of the incident and reflected ray for the hkl plane.

The intensity measured for a given reflection is proportional to F(hkl) where

$$F(hkl) = \sum_j f_j \exp\{2\pi i(hx_j + ky_j + lz_j)\}$$

where f_j is the atomic scattering factor for X-rays from the jth atom of coordinate (x_j, y_j, z_j) expressed as fractions of the cell a, b, c. This equation is the structure factor equation. The Fourier inverse of this equation is the electron density (ϱ) equation

$$\varrho(xyz) = 1/V \sum_h \sum_k \sum_l F(hkl) \exp\{\alpha(hkl)\} \exp\{-2\pi i(hx + ky + lz)\}$$

whereby if the amplitude and phase of the structure factor is known for all hkl planes or reflections, then the electron density can be calculated for all points (x, y, z) in the cell and so the crystal structure is then solved. Of course it is impossible to measure all h, k, l reflections so the summation is usually terminated with a finite number of terms at a certain resolution limit known as the Bragg resolution.

The problem of phase determination is the fundamental one in any crystal structure analysis. Classically protein crystallography has depended on the method of multiple isomorphous replacement (MIR) in structure determination. However lack of strict isomorphism between the native and derivative crystals and the existence of multiple or disordered sites limit the resolution to which useful phases may be calculated.

A separate source of information may be the anomalous scattering of atoms within the crystal. The atomic scattering is given by:

$$f = f + f' + if''$$

where f′ and f″ are the dispersion and absorbtion components of the anomalous scattering contribution to the total scattering factor. The presence of an f″ component in the total Bragg scattering leads to a breakdown of Friedel's law, i.e.:

$$I(hkl) \neq I(\bar{h}\bar{k}\bar{l})$$

which can be used not only to determine absolute configuration, but also phase angles (Bijvoet, 1949). The anomalous scattering of metals in protein heavy atom derivatives has been used to determine heavy atom positions by use of anomalous difference Pattersons (Rossmann, 1961), or to supplement isomorphous derivative information in phase determination (North, 1965; Mathews, 1966; Argos and Mathews, 1973). Anomalous scattering information alone may also be used in phase determination when only a single heavy atom derivative may be prepared or, more particularly, where an anomalously scattering atom is naturally present in the protein molecule. The method was first discussed by Herzenberg and Lau (1967), who suggested the use of naturally occurring sulphur within proteins; it has since been used successfully in phasing the structure of crambin (Hendrickson and Teeter, 1981). Smith and Hendrickson (1981) also used the method for phase determination of trimeric haemerythrin, where they took advantage of the anomalous scattering of the iron atoms, resolving the two-fold ambiguity in the phases, in both cases, by taking the phase closest to that of the anomalously scattering atoms.

The anomalous components of the total scattering are wavelength dependent and the use of radiation close to an absorption edge may increase or optimise the contribution due to the anomalously scattering atoms. Ramaseshan (1962) pointed out that data collected at multiple wavelengths optimising the anomalous dispersion effects would improve the quality of phase determination.

The availability of synchrotron sources with an almost continuous spectrum in the X-ray region allows the use of effects of metal atoms such as iron, zinc and copper which are often present in proteins as well as the atoms commonly used in derivative preparation, and a choice of the wavelengths at which the absorbtion edge is sampled.

Previously two wavelength experiments using conventional Ni and Cu target X-ray sources had been shown to be useful for the Fe edge in the pioneering work of Hoppe and Jakubowski (1975). Casscarano et al. (1982) extended the approach of Singh and Ramaseshan (1968) to show that two wavelengths, either both on one side or one either side of an absorption edge, direct methods may be used to determine the position of the anomalous scatterer. Karle (1980, 1982) has set out an algebraic analysis of the multiple wavelength experiment which allows the calculation of the amplitude of the contribution of the anomalous scatterers to the structure factor by the solution of linear equations involving two wavelength data, or by least squares analysis of data from three or more wavelengths. Woolfson (1984) has set out a proceedure for phase determination using anomalous scattering and Kahn et al. (1984) have obtained by an independent method an electron density map for a terbium derivative of parvalbumin using data collected at three wavelengths about the terbium absorbtion edge. A similar experiment carried out by Harada et al. (1987) used the anomalous scattering from the native haem iron atom in cytochrome c′. In all cases there was made the impicit assumption that the wavelength-independent component of the structure amplitude was known. Hendrickson (1984) has used the method of Karle (1980, 1982) cited earlier to obtain protein phases from multiple wavelength anomalous scattering data. The intensity changes in the diffraction pattern induced by anomalous contributions are small and require, usually, very accurate data. This new method in crystallography is being paralleled by theoretical developments in direct methods such as maximum entropy (Bricogne, 1984). Once protein phases are known anomalous scattering information may be used to obtain accurate metal-metal distances in proteins or with multiple wavelength data to distinguish between metal atoms of similar atomic weight within a protein. The first example of this involved the distinguishing between Ca and Mn in pea lectin (Einspahr et al. 1985). In an extended version of this experiment Kitigawa et al. (1987) unambiguously identified the Cu and Zn atoms in CuZn superoxide dismutase using data collected at five wavelengths, four about the Cu and Zn absorbtion edges, at the Photon Factory.

This chapter will concentrate on the uses of SR in protein crystallography involving the use of the high intensity and collimation for:

a) The reduction of radiation damage.
b) The study of large unit cells.
c) The use of small crystal volumes.
d) Kinetic crystallography, eg. of enzyme active states.
e) The study of molecular motion via diffuse scattering measurements with SR.

All the above techniques use incident monochromatic radiation, usually focussed in one or two dimensions. However for cases a) and d) the reduction of radiation damage and more particularly in kinetic crystallography the use of polychromatic data collection is yielding promising results. This technique makes combined use of the intensity and collimation of the SR beam with a large wavelength spread for Laue data collection from protein single crystals.

3 The Component Parts of the Experiment

The component elements of the protein crystallographic experiment are, briefly, as follows:

a) The X-ray Source.

The X-ray source may be a conventional sealed tube or rotating anode generator or bending magnet synchrotron radiation and more recently the exploitation of multipole insertion devices such as wigglers and undulators represent great gains in source intensity.

b) Beam Conditioning.

The insertion of conditioning materials in the X-ray beam alter and define the incident beam characteristics. i) Wavelength and wavelength spread determined using monochromators eg. mosaic graphite ($d\lambda/\lambda \approx 10^{-2}$) or Bragg reflection from a perfect crystal eg. Si or Ge ($d\lambda/\lambda \approx 10^{-4}$) either in a single or multiple crystal geometry possibly with cooling to avoid instabilities in wavelength due to beam heating effects on the monochromator. ii) Wavelength cut-off may be determined by critical reflection from a mirror giving a relatively sharp λ min limit, by transmission through a thin mirror or soap film giving a relatively sharp λ max limit and the insertion of metal foils to give a "soft" limit to λ max. iii) Beam crossfire or divergence may be controlled with slits. iv) Polarisation state of the beam may be controlled via the monochromator and choice of source. Mirrors and monochromators may also be used as focussing elements giving line or point focussed radiation and hence increased intensity at the sample.

c) The Sample Environment

The sample is placed in a controlled environment to define the temperature state of the sample with a cooling cell (Bartunik and Schubert, 1982, Helliwell, 1985), or the state of the mother liquor with a flow cell (Hajdu et al., 1985). Pressure cells have to data found little use in protein crystallography due to the fragility of most protein crystals, although Kundrot and Richards (1986) have collected data from lysozyme at high pressures to estimate the compressibility of protein structure and the effect of pressure on secondary structure and flexibility.

d) The Sample Holder.

The need for the protein crystal to be bathed in mother liquor is long established. Traditionally a glass capillary sealed with wax is used but work (Mahendrasingham, Sowerby and Helliwell, 1985 unpublished) at long wavelengths (2.6A) has led to the use of thin film capillaries (eg. Hostaphan, Hoescht Ltd.) since the thinnest glass capillaries absorb wavelengths around 2.6A by a factor of 20. These capillaries will give important benefits in protein microcrystal data collection.

e) The Detector.

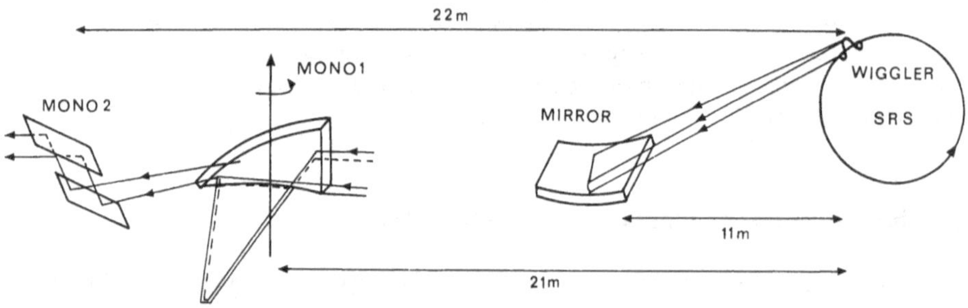

a

b

Fig. 1a and b. A schematic illustrating the mirror — monochromator layout for the wiggler protein crystallography worksattion at the SRS giving point focussed radiation at the sample (mono 1) or rapid tuneability (mono 2), the two options being mutually exclusive. **b** A view inside the protein crystallography experimental enclosure on the wiggler beam line at the SRS

The detector may be a single channel detector (scintillation counter on a diffractometer); film, imaging plate, MWPC, TV detector or CCD depending on the demands of the experiment eg. high count rate, high accuracy or good spatial resolution. Perhaps the most significant advance in detectors for crystallography recently has been the development of the re-useable, photostimulable imaging plate (Miyahara et al., 1986) with a much greater dynamic range, lower fog and greater sensitivity than film but otherwise with similar properties and applications. The characteristics and performance of detectors commonly used in diffraction experiments are discussed by Arndt (1987).

For a full discussion and review of instrumentation employed at SR centres see Helliwell (1984) and the references therein. In order to give an idea of the scale of the apparatus involved Fig. 1 shows a schematic of the optics on the SRS wiggler protein crystallography workstation and a view of the apparatus inside the experimental hutch.

4 Surveying Reciprocal Space in Quantitative Crystallography: Monochromatic and White Beam Laue Approaches

Traditionally quantitative X-ray crystal structure analysis is based on data collected by the monochromatic beam, rotating crystal method, conventionally using screenless oscillation camera geometry/diffractometry although some use has been made of a Weissenberg camera specifically designed for macromolecular crystallography (Arndt and Wonacott, 1977, Sakabe, 1983). Integration of reflection intensities is done over angle, known as the rocking width of each reflection. The value of the rocking width is determined by the intrinsic mosaic of the crystal, beam crossfire and spectral spread (Greenhough and Helliwell, 1982a, b). As the sample is rotated the leading and trailing edges of the Ewald sphere (of essentially single radius) sweep through reciprocal space. The optimum signal-to-noise ratio for a given reflection obviously occurs if the detector collects counts only during the angular rotation in the particular reflection rocking width.

Polychromatic techniques are finding increasing application at SR sources. The availability of synchrotron radiation X-ray sources has renewed interest in Laue diffraction methods which exploit directly the polychromatic nature of such sources. In the Laue method an incident beam with a wide range of wavelengths is used with a stationary crystal and detector, the integration of reflected intensities is done over wavelength. Preliminary studies of Laue diffraction from protein crystals (Moffat et al., 1984, 1986; Bilderback et al. 1984; Helliwell, 1984, 1985; Hedman et al. 1985) and small inorganic crystals (Wood et al., 1983; Hails et al., 1984) suggested that the Laue method posseses advantages over conventional monochromatic data collection strategies for certain experiments. It makes optimum use of the entire synchrotron radiation spectrum and the very high intensity of the white beam affords a reduction in exposure times of several orders of magnitude. The Laue method thus permits very brief exposure times in the millisecond time scale from a strongly scattering protein crystal (Moffat et al. 1984, 1986; Hajdu et al. 1986) and the examination of microcrystals (Hedman et al. 1985). The stationary crystal yields integrated intensities directly that are relatively insensitive to transient changes in unit cell dimensions or crystal orientation. A typical Laue diffraction contains many more reflections than a typical monochromatic photograph, corresponding to a much larger volume of reciprocal space being sampled. It can be shown that the equivalent rotation of the monochromatic case is about 15° of sample rotation. In favourable cases of high space group symmetry and careful choice of beam direction with respect to the crystal axes a single exposure may yield the bulk of the unique data (Elder, 1986). These advantages are particularly appropriate to dynamic experiments in which the diffracted intensities change rapidly with time in response to a structural perturbation with data collection rates approaching 30000 reflections per 0.25 second.

A fundamental complexity of the Laue method is the overlapping of many orders of each Bragg reflection which may be stimulated in a white X-ray beam and overlap exactly in scattering angle and poses a problem to Laue methods. Cruickshank, Helliwell and Moffat (1987) show that for a detector of unrestricted angular acceptance and even with an infinite wavelength spread that 72.8% of all Bragg reflections occur as singlet (spots consisting of a single reciprocal lattice point derived from a given wavelength in the available band pass). With more realistic experimental wavelength ranges this proportion increases to more than 83% and depends only on the ratio of the maximum to minimum wavelength incident at the sample and not on space group, crystal orientation or limiting resolution. The agreement of their theoretically derived distributions is very good. Hence multiple orders are not a serious limitation except for lower resolution reciprocal lattice points which tend to occur in multiplet spots. However, the modest energy resolution of X-ray film packs allows the resolution of constituent intensities in doublet Laue reflections.

A further complexity is the need to avoid or minimise spatial overlap of neighbouring reflections in often very dense diffraction patterns. This is usually minimised by using very small collimated beam sizes of the order of 0.2 mm diameter. This requirement does however improve the observed signal to noise characteristics of the Laue patterns, since the amount of extraneous material (eg. mother liquor, glass capillary) in the beam is limited the background scatter is reduced. Very short exposure times are preserved even with such a small incident beam and associated sample volumes because of the very high intensity (10^{14} polychromatic photons/sec/mm^2) in the beam compared with say the monochromatic focussed wiggler beam of 10^{12} photons/sec/mm^2 or a rotating anode of 10^9 CuK_α photons/sec/mm^2.

Why have Laue methods of surveying reciprocal space not been developed sooner? The power in the beam is high, as one can illustrate vividly if one places a highly absorbing sample such as lead in the SRS wiggler beam; it melts. Unfortunately, early experiments with the white beam on the NINA synchrotron with crystals of the enzyme 6-phosphogluconate dehydrogenase were totally unsuccessful (Bordas and Helliwell, 1976, unpublished, Clifton et al., 1985). Recently Moffat et al. (1984) used the so called modified Laue ($d\lambda/\lambda < 0.2$) with an SR beam from CHESS with successful results from myoglobin and haemoglobin crystals as a preliminary to laser photolysis studies on carbon monoxide debinding. Building on this work especially, Helliwell (1984) exposed a pea lectin crystal in the full ($0.2\ \text{A} < \lambda < 2.5\ \text{A}$) SRS wiggler white beam with beautiful results (see also Clifton et al. 1985). It seems to be the case that the more robust protein crystals (where robust here is defined as a property of relatively good lifetime in the monochromatic beam) survive several

(3–10) exposures whereas less robust crystals withstand a white beam of some limited wavelength range (modified Laue).

The interpretation of Laue patterns has required the development of new software for prediction, spot integration, film-to-film scale factor determination, harmonic spot unscrambling, the derivation of wavelength dependent correction factors and wavelength normalisation (Helliwell, 1985; Clifton et al. 1985; Campbell et al. 1986; Smith, Szebenyi, Schildkamp and Moffat, unpublished). The software is based on the already successful packages used for oscillation data processing (Arndt and Wonnacott, 1977; Rossmann et al. 1979; Vriend et al. 1986). The results from the pea lectin Laue data used as a test (Helliwell et al., 1986) are promising and compare favourably with monochromatic oscillation film data. Hence, data of useful statistical quality can be collected in very short times. This is of potential use for the study of protein crystal transient states. The technique is being actively developed in Cornell on CHESS and at Daresbury on the SRS.

5 Exploitation of the High Intensity and Superior Collimation of SR (with case Studies)

5.1 General

We can consider the total energy E(hkl) in a diffracted beam with reference to Darwins treatment for an ideally mosaic crystal rotating with constant angular velocity through the reflecting position.

$$E(hkl) = \frac{e^4}{m^2c^4\omega} I_0\lambda^3 PLA \frac{Vx}{V^2} |F(hkl)|^2$$

where λ is the wavelength of the incident beam of intensity I_0 the crystal volume is Vx, the unit cell volume V, P is a polarisation factor which depends on the state of polarisation of the incident beam and L is the Lorentz factor which takes into account the relative time each reflection spends in the reflecting position. From the equation we can see how, in principle, small crystal volume and large unit cell would reduce E(hkl) but be compensated for by an increase in I_0, the source intensity of a change in such as provided by an electron storage ring. From this equation we can clearly see how an intense SR source can be utilised; the high intensity can be used to overcome the following limitations:

(i) the inherent weakness of individual protein crystal reflections
(ii) the voluminous amount of data to be collected
(iii) radiation damage

Considerations (i) and (ii) are especially serious in the case of large unit cells. Small samples give weak diffraction patterns. All proteins suffer to a greater or lesser extent from radiation damage. Finally, problems (i) and (iii) affect the high resolution data the most; the fall off of the atomic scattering factor and atomic thermal vibration makes the high angle reflections weaker and radiation damage disrupts the diffraction pattern at atomic levels of resolution the most severely. In the following section we, discuss the use of SR in overcoming these problems. Additionally, new areas of protein structural research are opening up, for example, time resolved crystallography.

5.2 Reduction of Radiation Damage

That sample radiation damage should be reduced at an intense SR source seems perhaps somewhat paradoxical. However, phenomenologically there are important time dependent crystal processes involved in the radiation damage initiated by the exposure to the X-ray beam. These may be chemical/structural or lattice determining effects. Exposure times are reduced using SR to such an extent that these slower damaging processes can be ameliorated; this applies when increasing monochromatic or white beam intensities but eventually sample heating becomes limiting.

The reduction of radiation damage has several benefits. Firstly the initial resolution or maximum Bragg angle, that can be observed can seem better with SR. It is not that the data is absent in conventional source work, but that with SR statistically significant data can be collected before radiation decay occurs. The extension of data resolution may also be due, in part, to the well collimated geometry of the SR beam. Because of this the high angle reflections collected as spots on a film can be smaller than for a conventional source, thus making the average optical density recorded stronger and so more statistically significant. Secondly, more data per sample crystal can often be collected with SR to the extent that a single sample can yield a complete data set which was not previously routine.

An extreme example of the reduction in radiation damage is that of data collection at the SRS on purine nucleoside phosphorylase. On a conventional source usually a crystal can give only one 3 A resolution still photograph before the crystal suffers serious damage. At the synchrotron three crystals will give a complete set of 4 equivalent reflections to a resolution limit of 3 A (see case study below)

Case Study A: High Resolution Data Collection from 6-phosphogluconate Dehydrogenase (6-PGDH) Crystals.

The enzyme 6-phosphogluconate dehydrogenase is a dimeric protein of total molecular weight 100000 Daltons. The structure was solved at 2.6 A resolution using conventional X-ray sources and multiple isomorphous replacement (3 heavy atom derivatives) (Adams et al. 1983). Native data to 2 A has been collected using synchrotron X-radiation. This data has been processed with a merging R factor on intensity of 6% (Adams, M. J., pers. comm.), the data between 2 A and 2.6 A could not have been collected on a conventional source. This is due to the fact that this weak data would demand very long exposures on a conventional source such that radiation damage is too great. This is an example of time dependent radiation damage. The molecular model is now being refined at this higher resolution.

Case Study B: High Resolution Data from DPI on the FAST.

Data from DPI (beef des-pentapptide insulin) have been collected to 1.3 A resolution on the FAST (Fast Area Scanning Television) detector system (P. Holden, pers.

comm.). Several crystals would have been required to collect data to this resolution using film as the crystals were small ($0.15 \times 0.2 \times 0.3\ mm^3$) and require long exposure times leading to severe radiation damage. Data were collected from a single crystal (C2, a = 52.7 A, b = 26.2 A, c = 51.7 A, β = 93.4) as 0.2° rotation images to give a total of 180° about the b* axis, using a detector 2theta tilt of 11° and a crystal to

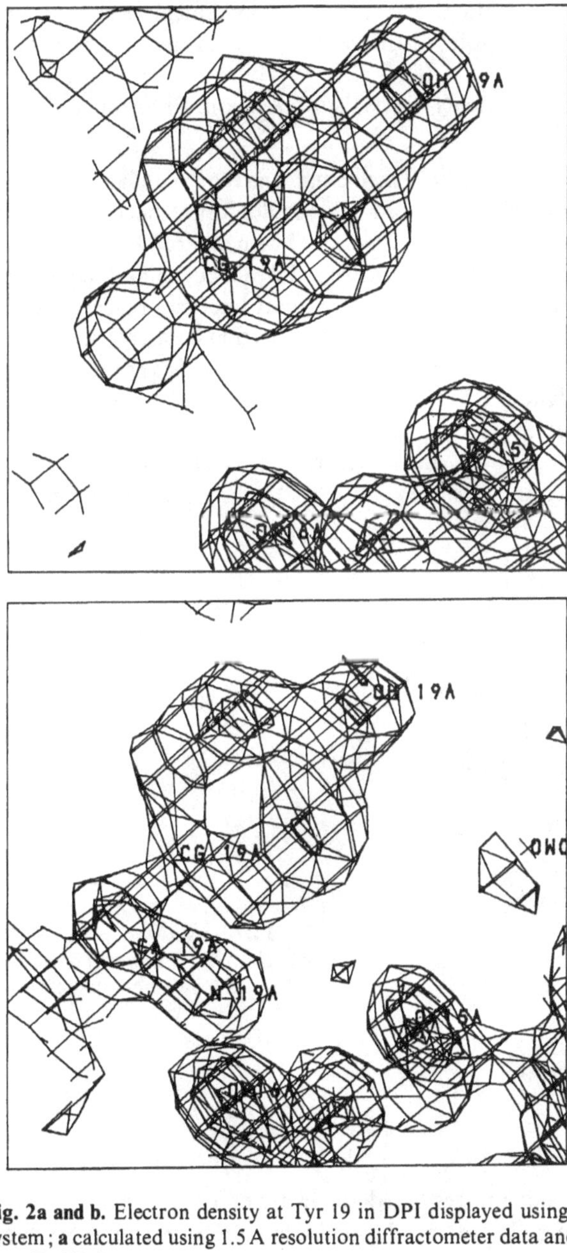

a

b

Fig. 2a and b. Electron density at Tyr 19 in DPI displayed using an Evans and Sutherland graphics system; **a** calculated using 1.5 A resolution diffractometer data and **b** calculated using 1.3 A resolution data FAST data detector distance of 44 mm. The data were processed using software developed by Pflugrath and Messerschmidt (1987) giving 14828 unique data at 1.3 A resolution and a merging R-factor of 11 %.

Figure 2 compares the electron density observed at Tyr 19 in the amino acid sequence calculated with data from diffractometer (at 1.5 A resolution) and the FAST. The structure factor phases were obtained from a preliminary refined structure and the density obtained from the FAST data shows a clear improvement over that from diffractometer as evidenced by the 'hole' in the centre of the aromatic ring density.

Case Study C: Large Solvent Content: Crystals of Human Erythrocyte Purine Nucleoside Phosphorylase.

This enzyme is of considerable biological, medicinal and commercial interest. It plays a fundamental role in purine metabolism, it degrades anti-cancer drugs being targetted to specific cancers and its absence is associated with severe T-cell deficiency. Crystals of the enzyme on a conventional source do not diffract beyond 4 A resolution, a resolution which is not adequate to study the structural interactions with various substrates and inhibitors. With intense synchrotron X-radiation data can be collected to 2.8 A resolution; this is sufficient to solve the structure. This is another example of time dependent radiation damage. Very fast data collection methods at the SRS involving collection of 100 reflections per second, where one crystal is used for 4 minutes of exposure time have allowed the structure to be solved at 3 A resolution (Fig. 3). Also, natural substrate and designed inhibitor binding can be studied.

Fig. 3. The structure of purine nucleoside phosphorylase determined using synchrotron radiation to a resolution limit of 3 A (crystals r32, a = 99.2 A = 92.1)

5.3 Large Unit Cells

The combination of all the SR advantages listed above are especially needed in virus crystallography where the unit cells are very large. Data collection from very large unit cell constants benefits from the natural narrow beam collimation of SR to resolve

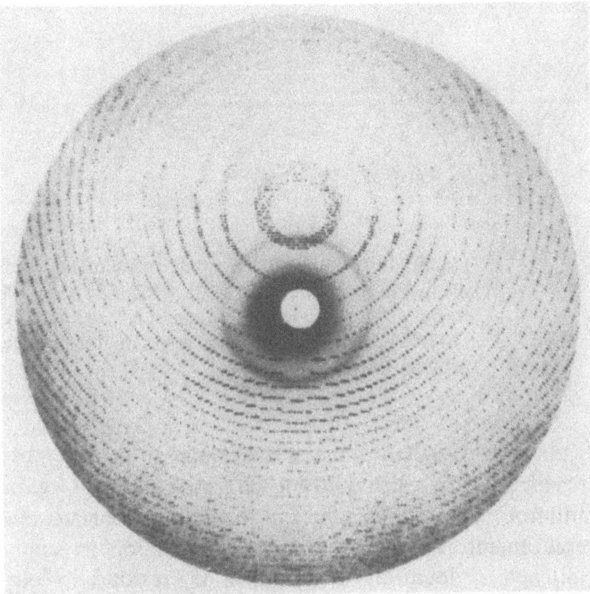

Fig. 4. 0.3° oscillation photograph from a crystal of cow pea mosaic virus taken with synchrotron radiation

adjacent spots (Fig. 4). Even with SR, exposure times can still be fairly long and in some cases only one data photograph can be collected per sample crystal due to radiation damage problems.

Further reductions in exposure time and hence radiation damage in virus crystallography may accrue from the use of white beam (modified) Laue methods; preliminary work on this is in progress (Bloomer and Helliwell (1985), unpublished at the SRS and Rossmann et al. at Cornell unpublished (1986)). Data collection on some virus crystals is virtually impossible in the home laboratory.

Case Study: Rhinovirus Data Collection

It took the short time of one year or so to solve the structure of rhinovirus which causes the common cold. This relied on two major advances in methods. The first was the use of synchrotron radiation in data collection. Nearly a million reflections were collected on the protein crystallography facility at the Cornell Synchrotron source in a matter of days. This conveyed a speed advantage over data collection on a conventional source and also ameliorated an otherwise impossible problem of radiation damage when long exposure times were used. The far greater rate of radiation damage in the X-ray beam in relation to plant viruses is symptomatic of an inherently less stable protein capsid and the absence of quasi-symmetry. The capsid consists of 60 copies each of four proteins and the virus with about 30% RNA has a total molecular weight of about 8.5 million.

The processing of the SR data by Rossmann and coworkers (Vreind et al., 1986) was specifically adapted to the treatment of SR data. Secondly, the intrinsic symmetry

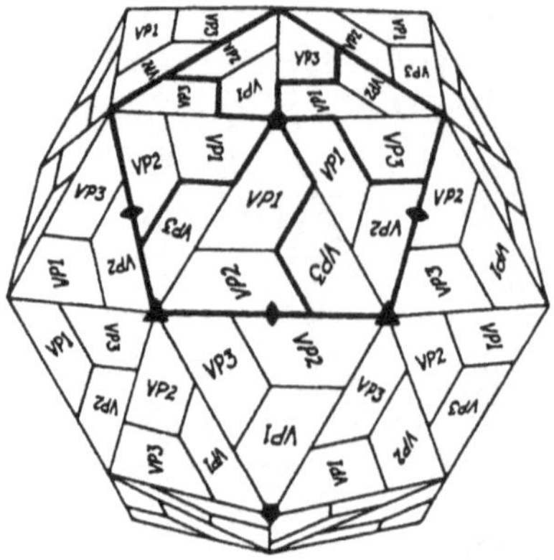

Fig. 5. Illustration of the arrangement of the four distinct polypeptide chains VP1, VP2, VP3 and VP4 forming the icosahedral protein shell surrounding an RNA core, the notation refers to the unique proteins of the virus shell

of the virus was used to provide phasing information in the final stages of the analysis, low resolution phasing was provided via two heavy atom derivative datasets again collected using synchrotron radiation. The high resolution phasing was carried out using real space molecular replacement whilst extending the resolution in steps. The final skew-averaged map allowed the location of 811 out of 855 residues in four distinct polypeptide chains (Fig. 5) (Arnold et al. 1987).

An important technical development is the use of short wavelengths in the study of the FMDV virus (Stuart and co-workers, unpublished) using the SRS wiggler; the combination of large unit cell and small crystal volume caused problems of signal to noise which were resolved by using a short wavelength of 0.9 A. The weaker diffraction being tolerated in order to reduce the background with long crystal to film distances.

The success of these new methods allows rapid structure determination of viruses. Moreover, since an intact animal virus can now be crystallised the surface proteins involved in immunogenic interactions can be viewed directly at the site of binding. This is very exciting and future studies will include other important pathogenic animal viruses and even more complex structure determinations of exceptional biological interest.

5.4 Small Sample Volumes

Several sample related quantities combine to affect the strength of the diffraction pattern in the following way

$$Vx \cdot 1/V^2 \cdot \text{Percentage solvent content}$$

where Vx = sample volume

V = unit cell volume

as seen above from Darwin's formula. High intensity is used to overcome this problem. Additionally to obtain a reasonable signal to noise ratio the X-ray background has to be minimised. This is especially true in the weak scattering case.

Many proteins currently selected for structural study crystallize only with small sample volume. This is especially true in the case of novel engineered mutants or

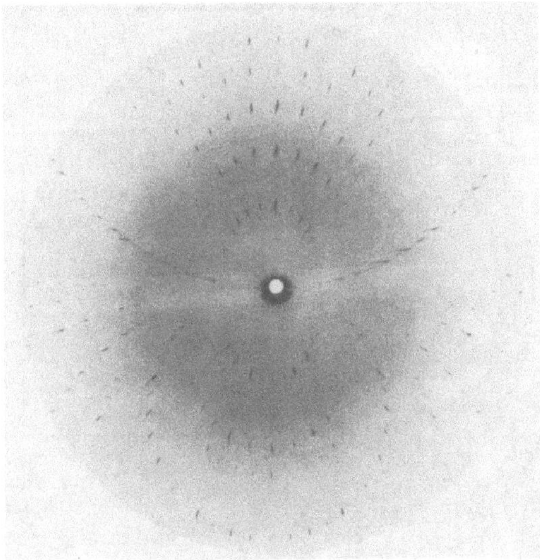

Fig. 6. Laue diffraction pattern recordered from a $30 \times 35 \times 10\ \mu m^3$ crystal of gramicidin A mounted in the usual way, sealed in a capillary with mother liquor. Exposure time 0.5 s

otherwise commercially interesting proteins, which are not readily crystallizable. The high intensity and collimation of the SR is currently allowing data to be collected and several structures to be studied where samples as small as $100 \times 100 \times 15\ \mu m^3$ and $200 \times 50 \times 50\ \mu m^3$ (Ealick, Helliwell and Cook, unpublished) with unit cell volumes in the range of 10^6 A^3 or so.. Sample volumes as small as $(20\ \mu m)^3$ have yielded very strong diffraction patterns with the white SR beam, (Fig. 6) (Hedman et al. 1985). In this application area efforts to reduce the X-ray background at the detector include replacement of traditional glass capillaries in the sample with thin film capillaries.

Andrews et al. (1986) have exploited the advantages of SR using the protein crystallography work station at the SRS to solve the structure of a silicate using monochromatic data collected from a crystal of $18 \times 8 \times 175\ \mu m^3$. The crystal was mounted dry, in air, data collected using a collimated beam of 0.2 mm diameter wavelength of 0.88 A to improve signal to noise and the Enraf-Nonius FAST television area detector. The refined mosaic spread was high (2–3°), typical of such small samples and the processed data of relatively low quality, in particular the weak data. The structure however was solved using direct methods and refined to a crystallographic R value of 10.3%. Although a small molecule the result is an encouraging example of what can be achieved in the field of data collection from extremely small sample volumes, impossible to contemplate using laboratory X-ray sources.

5.5 Time Resolved Crystallography

The intensity of SR is high enough for crystallographic data to be collected in real time to give direct time resolution of dynamic events in a protein molecule. Bartunik et al. (1982) collected monochromatic test data on a millisecond time scale on carbonmonoxy myoglobin, where structural changes are induced by the debinding of the

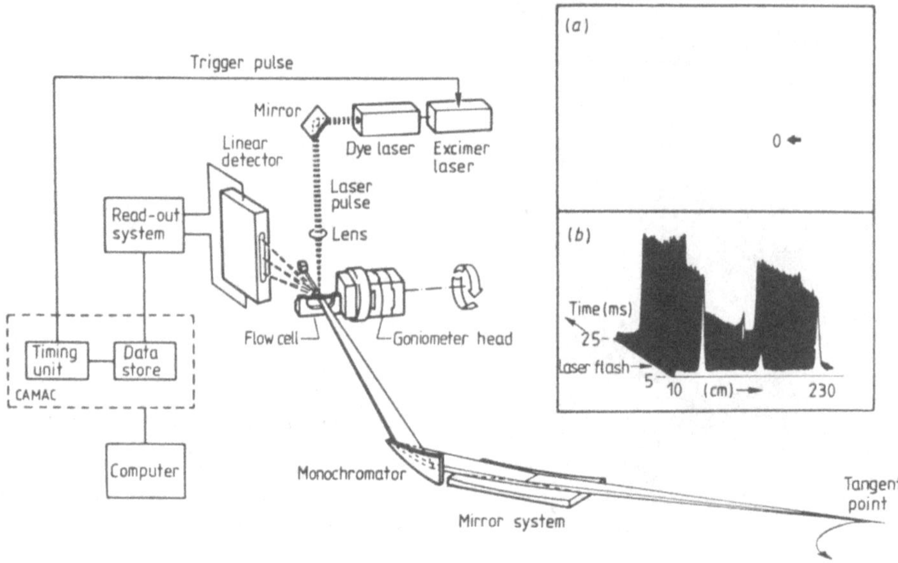

Fig. 7. The instrumental set up at X11/DORIS for time resolved data collection with a linear detector for CO myoglobin following laser photolysis of the ligand. A section of the diffraction pattern with stationary crystal, stationary detector is recorded with a linear detector and (**b**) shows the time course of three reflections before and after the laser flash (from Bartunik et al. 1982)

CO ligand by a laser (Fig. 7). As mentioned earlier Laue methods have been used which avoid the effects of non-isomorphism in a kinetic sequence.

Although in principle the time structure of the SR beam may be exploited in time resolved studies the major limiting factor is the rate at which three dimensional data may be accumulated. In this respect time resolved methods are bound to develop in tandem with the development of high count rate/fast refresh rate electronic area detectors. This applies to both monochromatic and white beam methods. For the latter the use of an integrating detector such as a CCD or image plate are the main expected improvements over film.

Case Study: Kinetic Crystallographic Studies of the Phosphorylase Enzyme.

The structural study of enzyme function in general would, ideally, involve the direct imaging in an electron density map of the enzyme substrate complexes. However, the lifetime is relatively short for such a complex. The enzyme turnover rate defines the number of substrate molecules converted per second and varies, i.e. for different enzymes, over a range from 10^7 per second to 2 per second at room temperature in solution. To increase the lifetime of the complex sufficiently for data collection on a low intensity source, cooling is needed to around -100 °C. However many protein crystals are grown from an aqueous medium, the crystals themselves contain large solvent channels and hence on cooling below zero the crystals fracture. It is sometimes possible to modify the aqueous solvent to include "anti-freeze" agents. In many interesting cases this is not feasible without chemically destabilising the crystal lattice. It becomes clear therefore that there is considerable scope for the use of high intensities and rapid data collection techniques in the study of enzyme transition states whilst avoiding extreme cooling of the sample. This approach has been used for the first time by the group at Oxford University led by Dr. L. N. Johnson studying the enzyme glycogen phosphorylase-b (Hajdu et al., 1987).

The experiments have used both monochromatic and white beam methods. We shall summarise both approaches starting with the monochromatic data collection experiments.

Monochromatic Experiments

The first experiment involved the substrates heptenitol and phosphate, where a long incubation time (50 h) in the crystal gave rise to a product, heptulose-2-phosphate, bound at the active site. Recently the group have been able to trap the actual enzyme-substrate complex (McLaughlin et al. 1984). A 1 h soak followed by 2 h data collection on the SRS (over 100000 reflections) showed a mixture of substrate and product. A 15 minute soak followed by 45 minute data collection at room temperature resulted in an electron density map which clearly showed substrate (phosphate and heptenitol) and no product bound. Thus, for the first time, by taking advantage of the very rapid methods of data collection (especially on the wiggler station), both the enzyme-substrate complex and enzyme-product complex were trapped under conditions in which the enzyme is fully competent.

Further experiments with the substrate maltoheptose and inorganic phosphate showed that this reaction was faster than with heptenitol. A 15 minute soak followed by data collection at room temperature resulted in the formation of the product, glucose-1-phosphate at the active site. At lower temperature (3 °C) with only a 5 minute soak of phosphate with crystals already pre-equilibrated in maltoheptose and a 35 minute data collection the reaction was found to be stopped. Hence with only a small adjustment in temperature and in a range where the crystals are stable it was feasible to explore the reaction pathway by crystallographic methods on this

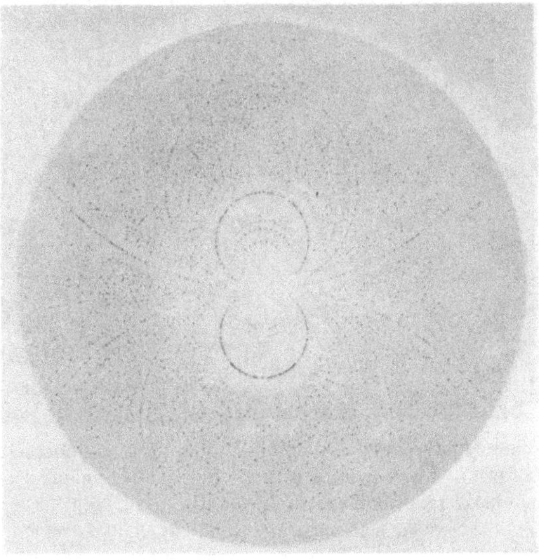

Fig. 8. A full white beam Laue diffraction pattern recorded from a crystal of glycogen phosphorylase b, and (P4(3)2(1)2 a = b = 128.8 A, c = 116.2 A) using a 0.2 mm diameter collimator the exposure time was 0.5 s

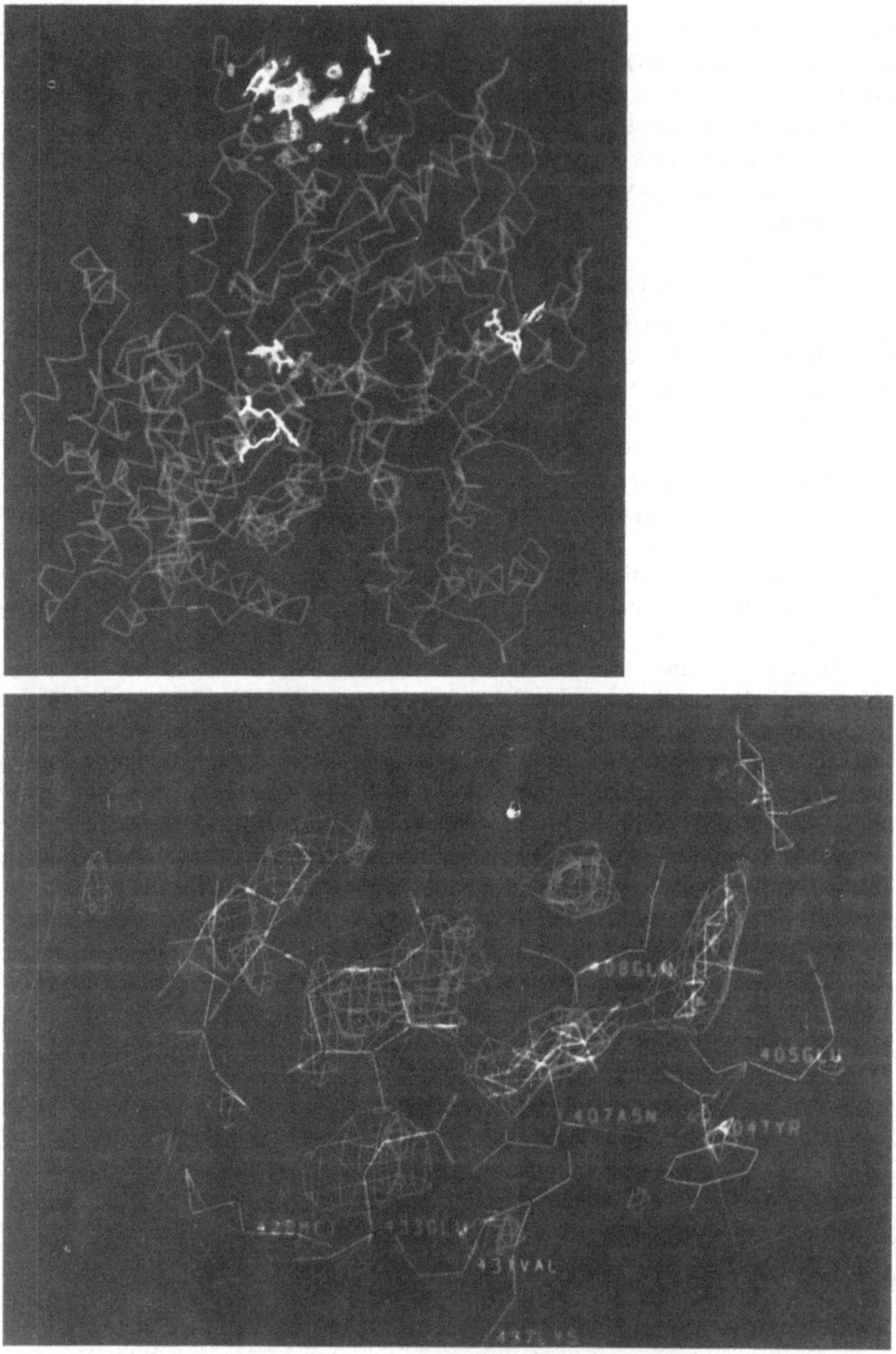

Fig. 9a and b. Difference Fourier maps calculated from Laue diffraction data showing maltoheptose bound in phosphorylase **b**. The Laue map shown in **a** is calculated with a subset of 9029 unique data at 2.5 A resolution. A positive contour at half maximal peak height is shown. **b** is an enlargement of **a** and shows 4 of the 7 sugar units, the 3 central units have the highest occupancies. Side chain movements produce the two extra lobes of density. (Figures courtesy of J. Hajdu)

enzyme (Hajdu et al., 1987). With the focussed monochromatic beam used at that time on the wiggler line of the SRS the intensity level was 10^3 higher than a rotating anode source. The 35 minute data collection time referred to above involved about 10 mins exposure time and 25 minutes of film cassette changes and represents a reflection measuring rate from the crystals during exposure of approximately 200 per second. Improvements currently underway at the SRS will reduce data collection times to a few minutes in total.

Full White Beam Laue Data Collection

Another approach to these experiments is also being followed, which involves the use of the full white beam from the SRS wiggler. The geometry of the Laue method is advantageous since a sir.gle exposure (Fig. 8) yields 30% of all data to 3 A resolution for the tetragonal crystal form of phosphorylase and exposure times in the millisecond time scale.

In the Laue experiment the binding of maltoheptose to the glycogen storage site of phosphorylase was monitored. The crystal was kept in a thermostatted flow cell and Laue datasets of three exposures (6 films in a pack) of 1 second were taken at different angular settings before, during and after the addition of maltoheptose. Each film in a film pack was scaled to its counterpart in the native Laue dataset and the measurements kept unmerged. The initial photographs represent the native protein and the spectral properties of the X-ray beam were not altered, fractional intensity changes on subsequent photographs with the same crystal orientation may be used to calculate a differencc courrier map with the coefficients

$$\frac{(\text{F-laue-der} - \text{F-laue-native})}{\text{F-laue-native}} \times \text{F-mono-native}$$

and phases from the monochromatic data, where F-mono-native is the measurement from a reference native dataset, F-laue-native the structure factor amplitude measured on the starting native photograph and F-laue-deri the corresponding measurement after addition of the ligand. The difference Fourier map showing maltoheptose at the glycogen storage is shown in Fig. 9 is of better quality than the equivalent monochromatic map (calculated with identical subsets of data) and clearly show 4 of the 7 sugar units (three protrude into solvent and show increased mobility) and difference density due to side chain movements in the active site (J. Hajdu, L. N. Johnson and co-workers 1988. It is the first example of an electron density map from a protein crystal using the Laue method of data collection and used a total data collection time of only 3 seconds per dataset.

The advantages of the difference Laue technique is that wavelength and position dependent corrections are not needed except in the case of significant change in the mass absorbtion coefficient during the course of the experiment (eg. the binding of heavy atoms). A modified difference Laue method was exploited by Faber et al. (1987) in the determination of the positions of the heavy atom binding sites in crystals of xylose isomerase.

It is possible to extract structural information from Laue data without recourse to a starting, native structure. The success of this approach does however depend on a satisfactory treatment of the wavelength dependent effects such as incident intensity, absorbtion, Lorentz and polarisation factors, obliquity and detector response and the determination of the "wavelength normalisation curve" to compensate for these effects.

The software required for the quantitative evaluation and wavelength normalisation of Laue data has been successfully developed at Daresbury since the feasibility of recording full white beam Laue patterns from protein crystals was established. The software and the method have been tested using Laue data from crystals of pea lectin. As an example of the statistical quality (to 3 A resolution) the mean fractional change on F between monochromatic and Laue pea lectin data was 11% and the same quantity between conventional source monochromatic and SR monochromatic pea lectin data is 8% (Helliwell et al., 1986).

5.6 SR and Diffuse Scattering from Protein Crystals: Molecular and Lattice Dynamics

Of great interest to the molecular biologist is the relationship of protein form to function. Recent years have shown that although structural information is necessary, some appreciation of the molecular flexibility and dynamics is essential. Classically this information has been derived from the crystallographic atomic thermal parameters and more recently from molecular dynamics simulations (see for example McCammon 1984) which yield independent atomic trajectories. A characteristic feature of protein crystals, however, is that their diffraction patterns extend to quite limited resolution even employing SR. This lack of resolution is especially apparent in medium to large proteins where diffraction data may extend to only 2 A or worse, thus limiting any analysis of the protein conformational flexibility from refined atomic thermal parameters. It is precisely these crystals where flexibility is likely to be important in the protein function.

Diffuse scattering, which nearly always occurs during a diffraction experiment however, represents a potentially rich source of dynamic information supplementary to that obtainable from the Bragg reflections. Many protein crystals exhibit diffuse

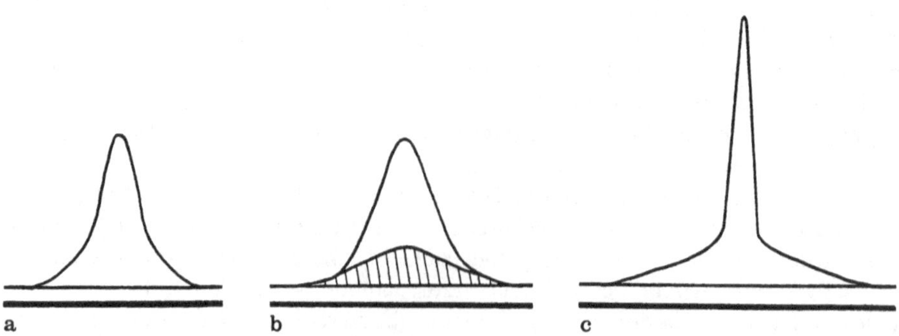

Fig. 10a–c. Schematic diagram illustaring reflection profiles for **a** an ideal Einstein solid and **b** a real solid, showing the contribution to the reflection of the acoustic diffuse scattering component. **c** illustrates the use of very fine collimation to distinguish the Bragg peak from the broader diffuse scattering shoulder, minimising the diffuse scattering contribution to the integrated intensity

scattering and this scattering is strongest in those crystals where Bragg reflections are few and atomic displacements are greatest whereby crystal disorder or lattice vibrations are large.

The acoustic diffuse scattering, arising from long wavelength acoustic vibrations, peaks at the Bragg positions (Fig. 10) and unless correctly measured or corrected for, represents a source of error in intensity measurements. The molecular diffuse scattering, present between the Bragg peaks is heterogeneous and direct evidence of correlated atomic displacements taking place in the crystals. Many motions taking place are

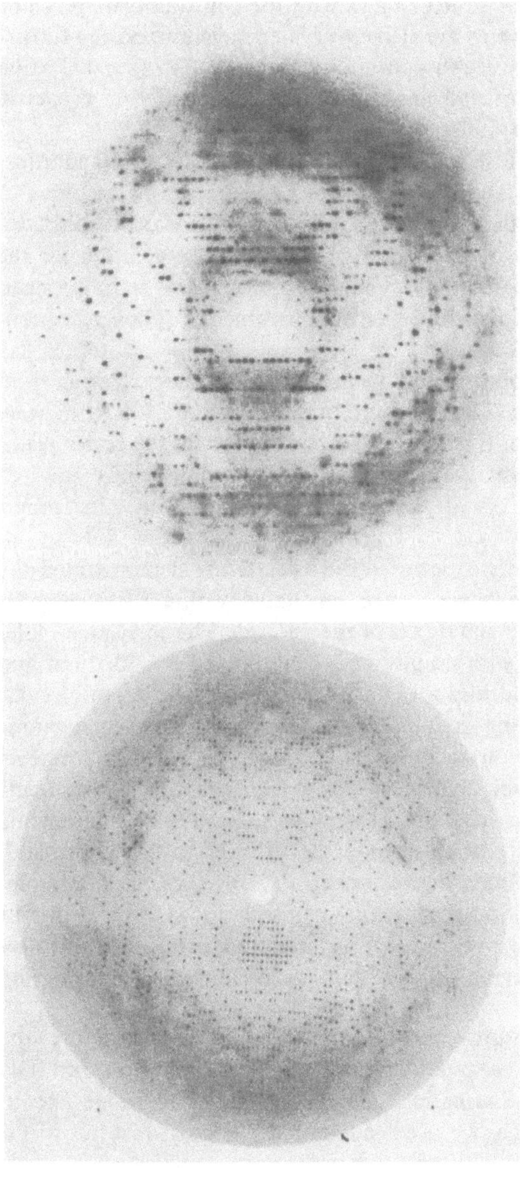

a

b

Fig. 11a and b. Oscillation patterns recorded from **a** 6-phosphogluconate dehydrogenase and **b** rabbit serum transferrin showing rich and highly oriented molecular diffuse scattering features

believed to involve concerted motion of groups of atoms which may be side groups, helices, larger groups of secondary structure or even domains moving about a hinge and intermolecular correlated displacements.

The measurement of diffuse scattering imposes severe experimental conditions for data collection. Firstly the separation of Bragg from acoustic scattering requires extremely fine collimation, producing a very fine Bragg peak against the broader shoulder of diffuse scattering (Fig. 10). The very fine collimation and minimal beam divergence with useable intensity requires the use of SR.

The molecular diffuse scattering due to its origins in inelastic events is of very much lower intensity than the Bragg data and to be collected in reasonable time scales with good statistics requires a very intense source, again with the constraints of very fine collimation. The diffuse scattering events are also energy dependent, imposing further conditions on the spectral divergence. All conditions make synchrotron radiation sources prerequisite in the treatment and analysis of inelastic scattering events in protein or other macromolecular crystals.

Using the SRS highly oriented diffuse scattering patterns from 6-PGDH and transferrin have been collected (Fig. 11) and the intensities being processed in terms of a series of non-integral reciprocal lattices for use in a refinement of correlated displacements in the protein crystals. We are also deriving an empirical correction for the acoustic scattering components to correct a very high resolution data set in order to assess the effects on refined positional and thermal parameters (Glover, Harris, Helliwell and Moss (1986) unpublished).

Doucet and Benoit (1987) have collected and measured the diffuse scattering from the orthorhombic form of lysozyme. The exposure times using the LURE SR were 30 minutes compared to exposures of the order of seconds to record the Bragg peaks alone. The diffuse scattering features they observed were predominantly rows of modulated intensity along two families of planes (100) and (001) with other minor contributors.

Their analysis of the measured intensities of diffuse scattering demonstrated the existence of rigid body displacements which are correlated within short periodic rows of aligned molecules along the a and c axes of the unit cell. The correlated molecules form two types of super-unit with amplitudes of motion of 5×10^{-4} nm and 7×10^{-4} nm, within which the molecules are sterically coupled via intermolecular interactions. Where no steric coupling or interactions are observed no correlations in displacement were observed. The physical origins of the correlated displacements were assigned to local disorder in one or more of the residues involved in contacts with a neighbouring molecule producing a displacement in the latter, all displacements being correlated within a superunit. In this study the diffuse scattering provides direct evidence only about the rigid body displacements of molecules, i.e. intermolecular motions and contacts whereas other proteins may well show features due to intramolecular motions. It does however represent an initial experiment in this new approach to protein dynamics deriving information inaccesible when considering Bragg scattering alone.

An alternative approach uses intuitive arguments about protein structure and flexibility using molecular graphics. A putative hinge region is present in 6-PGDH relating domains of the structure. Simulation of motion about this hinge and its effect on the molecular transform may be used to simulate the gross features of the molecular diffuse scattering Hellwell et al. 1986. This may be extended, on smaller molecules, to a calculation of the low frequency normal modes of vibration which may be used, via calculation of the transforms, in simulations of the recorded molecular diffuse scattering.

6 Conclusions and Future Directions

Protein crystallography is now a well established part of the research programmes of the synchrotron radiation sources whose spectrum extends into the X-ray region. SR considerably increases the rate at which data are collected. The higher resolution data that are measured improves the refinement of the macromolecular structures which is most important in understanding functions such as catalysis. The development of intense radiation sources by means of wiggler insertion devices permits the investigation of weakly diffracting crystals whether this is due to the size of the crystal (microcrystals) or due to the large size of the unit cell or some disorder or large solvent content of the crystal. The technique of optimised anomalous dispersion coupled to an intense source with tuneable wavelengths in the range 0.2–3.5 A may become a more direct method of solving the crystallographic phase problem for small and medium sized protein single crystal structures than those used at present.

The SR spectrum of an intense energy continuum of radiation makes possible rapid data collection in the sub-second regime by the Laue diffraction method. This rate of data collection is sufficient to investigate some of the crystal kinetics and catalytic intermediate reaction states that can be produced.

X-ray undulator devices and new high brightness multipole wiggler magnets offer to considerably extend the range of kinetic crystallography into the micro-second regime and less and small sample volumes of less than $(20\ \mu m)^3$ for units cells $(100\ A)^3$ upwards. Dedicated sources of this type have been proposed, namely the European Synchrotron Radiation Source and the USA 6GeV ring. The first X-ray undulator beam line becomes available on PEP in Stanford in 1987 for parasitic use.

7 Acknowledgements

We are grateful to many colleagues using SR at various facilities around the world for many useful discussions and for kindly supplying details of their work. For the work directly involving the authors, facilities were provided by the Universities of York and Keele, Birkbeck College and the Science and Engineering Research Council's (SERC) Daresbury laboratory. The SERC is thanked for research grant support. JRH also wishes to acknowledge the support provided by EMBO, NATO and the Royal Society.

8 References

Adams, M. J., Archibald, I. G., Bugg, C. E., Carne, A., Gover, S., Helliwell, J. R., Pickersgill, R. W., White, S. W.: EMBO J. *2*, 1009–1014 (1983)

Andrews, S. J., Papiz, M. Z., McMeeking R., Blake, A. J., Lowe, B. M., Franklin, K. R., Helliwell, J. R., Harding, M. M.: Acta Cryst *B44* 73–77 (1988)

Argos, P., Mathews, F. S.: Acta. Cryst. *B29*, 1604–1611 (1973)
Arndt, U. W.: J. Appl. Cryst. *19*, 145–163 (1986)
Arndt, U. W., Wonacott, A.: in: The Rotation Method in Crystallography (Arndt and Wonacott eds.), North Holland Publishing Co., Amsterdam (1977)
Arnold, E., Vreind, G., Luo, M., Griffith, J. P., Kamer, G., Erickson, J. W., Johnson, J. E., Rossmann, M. G.: Acta. Cryst. *A43*, 346–361 (1987)
Bartunik, H. D., Fourme, R., Phillips, J. C.: in: Uses of Synchrotron Radiation in Biology (Ed. Sturhmann, H. B.), Academic Press New York (1982)
Bartunik, H. D., Schubert, P.: J. Appl. Cryst. *15*, 227–231 (1982)
Bilderback, D. H., Moffat, K., Szebenyi, D. M. E.: Nucl. Instrum. Meth. *222*, 245–251 (1984)
Bijvoet, J. M.: Proc. Acad. Sci. Amst. *52*, 313 (1949)
Bricogne, G.: Acta. Cryst. *A40*, 410–445 (1984)
Campbell, J., Habash, J., Helliwell, J. R., Moffat, K.: Inf. Quat. Prot. Cryst. Daresbury Lab. *18*, 35–38 (1986)
Cascarrano, G., Giacovazzo, G., Pendeman, A. F., Kroon, J.: Acta. Cryst. *a38*, 710–717 (1982)
Clifton, I. J., Cruickshank, D. W. J., Diakun, G., Elder, M., Habash, J., Helliwell, J. R., Liddington, R. C., Machin, P. A., Papiz, M. Z.: J. Appl. Cryst. *18*, 296–300 (1985)
Cruikshank, D. W. J., Helliwell, J. R., Moffat, K.: Acta. Cryst. in press (1987)
Doucet, T., Benoit, J. P.: Nature *325*, 643–645 (1987)
Einspahr, H., Suguna, K., Suddath, F. L., Ellis, G., Helliwell, J. R., Papiz, M. Z.: Acta. Cryst. *B41*, 336–341 (1985)
Elder, M.: Inf. Quat. Prot. Cryst. Daresbury Lab. *19*, 31–37 (1986)
Faber, G. K., Machin, P. A., Almo, S., Hajdu, J., Petsko, G. A.: Proc. Natl. Acad. Sci. USA in press (1987)
Greenhough, T. J., Helliwell, J. R.: J. Appl. Cryst. *15*, 338–351 (1982a)
Greenhough, T. J., Halliwell, J. R.: J. Appl. Cryst. *15*, 493–508 (1982b)
Greenhough, T. J., Helliwell, J. R.: Prog. Biophys. Molec. Biol. *41*, 67–123 (1983)
Hails, J., Harding, M. M., Helliwell, J. R., Liddington, R. C., Papiz, M. Z.: Daresbury Lab. Preprint DL/SCI 479E (1984)
Hajdu, J.: Inf. Quat. Prot. Cryst. Daresbury Lab. *17*, 17–22 (1986)
Hajdu, J., Acharaya, K. R., Stuart, D. I., McLaughlin, P. J., Barford, D., Oikonomakos, N. G., Klein, H., Johnson, L. N.: EMBO J. *6*, 539–546 (1987)
Hajdu, J., McLaughlin, P. J., Helliwell, J. R., Sheldon, J., Thompson, A. W.: J. Appl. Cryst. *18*, 528–532 (1985)
Hajdu, J., Machin, P. A., Campbell, J. W., Greenhough, T. J., Clifton, I. J., Zurek, S., Gover, S., Johnson, L. N., Elder, M.: Nature *329* 178–181
Harada, S., Masanori, Y., Murakawa, Y., Kasai, N., Satow, Y.: J. Appl. Cryst. *19*, 448–452 (1986)
Hedman, B., Hodgson, K. O., Helliwell, J. R., Liddington, R. C., Papiz, M. Z.: Proc. Natl. Acad. Sci. USA *82*, 7604–7607 (1985)
Helliwell, J. R.: Rep. Prog. Phys. *47*, 1468–1497 (1984)
Helliwell, J. R., Glover, I. D., Jones, A., Pantos, E., Moss. D. S.: Biochem. Soc. Trans. *14*, 653–656 (1986)
Helliwell, J. R., Papiz, M. Z., Glover, I. D., Habash, J., Thompson, A. W., Moore, P. R., Harris, N., Croft, D., Pantos, E.: Nucl. Instrum. Meth. *A246*, 617–623 (1984)
Hendrickson, W. A.: I.U.Cr. 13th Int. Congress Abstr. ML13-H4 (1984)
Hendrickson, W. A., Teeter, M. M.: Nature *290*, 107–113 (1981)
Herzenberg, A., Lau, M. S. M.: Acta. Cryst. *22*, 24–28 (1967)
Hoppe, W., Jakubowski, U.: in: Anomalous Scattering (Ed. Ramaseshan, J., Abrahams, S. C.), Munksgaard (1975)
Kahn, R., Fourme, R., Bosshard, R., Chiadmi, M., Risler, J. L., Dideberg, O., Wery, J. P.: FEBS Letts *179*, 133–137 (1985)
Karle, J.: Int. J. Quantum. Chem. *7*, 357–367 (1980)
Karle, J.: in: Computational Crystallography (Ed. Sayre) (1982)
Kitigawa, Y., Tanaka, N., Hata, Y., Katsube, Y., Satow, Y.: Acta. Cryst. *B43*, 272–275 (1987)
Kundrot, C. E., Richards, F. M.: J. Appl. Cryst., 208–213 (1986)
Matthews, B. W.: Acta. Cryst. *20*, 82–86 (1966)
McCammon, J. A.: Rep. Prog. Phys. *47*, 1–46 (1984)

McLaughlin, P. J., Stuart, D. I., Klein, H. W., Oikonomakos, N. G., Johnson, L. N.: Biochemistry *23*, 5862–5873 (1984)
Miyahara, J., Takahashi, K., Amemiya, Y., Kimiya, N., Satow, Y.: Nucl. Instrum. Meth. *A 246*, 572–578 (1986)
Moffat, K., Schildkamp, W., Bilderback, D. H., Volz, K.: Nucl. Instrum. Meth. *A 246*, 617–623 (1986)
Moffat, K., Szebenyi, D., Bilderback, D.: Science *223*, 1423–1425 (1984)
North, A. C. T.: Acta. Cryst. *18*, 212–216 (1965)
Messerschmidt, A., Pflugrath, J. W.: J. Appl. Cryst. *20*, 306–315 (1987)
Ramaseshan, S.: in: Advanced Methods in X-ray Crystallography (Ed. Ramachandran, G. N.), Academic Press (1962)
Rossmann, M. G.: Acta. Cryst. *14*, 383–388 (1961)
Rossmann, M. G.: J. Appl. Cryst. *12*, 225–238 (1979)
Sakabe, N.: J. Appl. Cryst. *16*, 542–547 (1984)
Vreind, G., Rossmann, M. G., Arnold, E., Luo, M., Griffith, J. P., Moffat, K.: J. Appl. Cryst. *19*, 134–139 (1986)
Wood, I. G., Thompson, P., Matthewman, J. C.: Acta. Cryst. *B39*, 543–547 (1983)
Woolfson, M. M.: Acta. Cryst. *A 40*, 32–34 (1984)

Synchrotron Light on Ribosomes: The Development of Crystallographic Studies of Bacterial Ribosomal Particles

Klaus S. Bartels[1], Gabriela Weber[1], Shulamit Weinstein[2], Heinz-Günter Wittmann[3], and Ada Yonath[1,2]

Table of Contents

[1] Max-Planck-Arbeitsgruppen für strukturelle Molekularbiologie, clo DESY, Notkestr. 85, D-2000 Hamburg 52
[2] Weizmann Institute of Science, Rehovot, Israel
[3] Max-Planck-Institut für molekulare Genetik, Ihnestraße 73, D-1000 Berlin 33

Topics in Current Chemistry, Vol. 147

1 Introduction

If ever a desperate project benefited from the availability of synchrotron radiation, that project is the crystallographic study of ribosomal particles.

Ribosomes are unique assemblies of several strands of RNA and of a large number of different proteins, representing the living cell's protein factory. Upon initiation of the biosynthetic process, one larger and one smaller subunit associate to form the active cell organelle that reads genetic information from the messenger RNA and translates it into a specific polypeptide chain. To this end, several binding sites are provided: for the messenger, for the tRNAs carrying the amino acids, for GTP, and a variety of other factors.

The large subunit of a typical eubacterial ribosome (MW ca. 1,600,000) is composed of 2 RNA chains and about 35 different proteins, the small subunit (MW ca. 700,000) comprises 1 RNA chain and about 21 proteins.

The chemical and physical properties of ribosomes are well characterized (for reviews see [1-5]). The exact understanding of their function, however, still lacks a detailed molecular model. Appropriate methods such as image reconstruction from electron micrographs of two-dimensional sheets, or X-ray structure analysis, all depend on the crystallizability of the material.

Experiments towards growing ribosomal crystals in vitro were challenged by the observations that ribosomes may self-organize into ordered aggregates in the living cell; two-dimensional arrays have been found under special conditions, such as hibernation or lack of oxygen (e.g. [6-8]). However, the complex structure, the enormous size and the flexibility of ribosomal particles render their crystallization in vitro extremely difficult. Therefore, only a few successful efforts to produce three-dimensional crystals have been reported (e.g. [9-11]).

2 From "Powder" Samples to Single Crystals — Some History

For a number of years now we have been involved in crystallization of bacterial ribosomal particles. From the very beginning of our studies, the crucial need for a stable, very intense, and perfectly focussed synchrotron beam was evident, even for preliminary and basic information (e.g. whether crystals diffract at all). Thus our studies have always been dependent on the availability of synchrotron beam time and hampered by only partial (and very occasional) feedback to assess our experimental procedures for growing bacteria, preparing the ribosomes and obtaining crystals.

2.1 E. coli and B. Stearothermophilus

Our first microcrystals were of Bacillus stearothermophilus 50S subunits [9] and of 70S ribosomes from E. coli [12]; they were mainly grown from lower alcohols, toluene, or chloroform. Each ribosomal preparation required slightly different crystallization conditions, and often the preparation had almost been exhausted by the time conditions were optimised. We also found that crystals grew from active particles only.

Every step in our crystallization experiments was checked carefully with electron microscopy and a high magnification light microscope (100× and more); lower magnification would often not reveal the minute first signs of success.

Large amounts of microcrystals from B. stearothermophilus 50S subunits were collected in X-ray capillaries (and even larger amounts were lost during this procedure), centrifuged to achieve dense packing, and carried to the X-ray beam. Imagine the atmosphere of a synchrotron concrete bunker and the excitement in the middle of the night, when — for the first time — the small-angle X-ray photograph of a supposed suspension of microcrystals really showed Debye-Scherrer diffraction rings of a "powder" sample, albeit at very low Bragg resolution!

This stage of our studies can be described as endless attempts to obtain meaningful diffraction patterns. High angle diffraction photographs would occasionally show weak but sharp rings (Fig. 1a), some of them with spacings similar to those that were previously reported for gels of ribosomes and extracted rRNA (e.g. 10.4 Å, 7.8 Å, 5.6 Å, 4.9 Å, and 3.4 Å; cf. [13–15]).

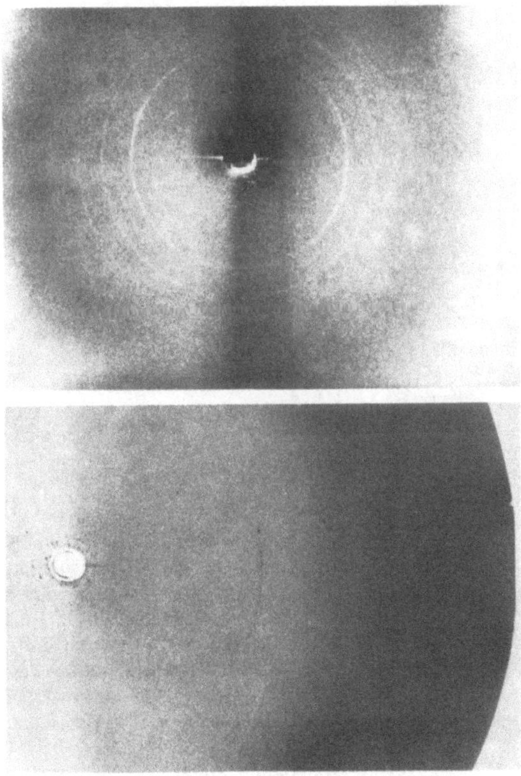

a

b

Fig. 1a and b. Early diffraction patterns (obtained at X11/EMBL/DESY) show few reflexion spots from minicrystals and fiber diffraction from the RNA moiety. **a** Unordered crystals; several full rings with spacings that arise from RNA; **b** partly ordered crystals, as described in the text, two arcs of fiber diffraction

Later on we succeeded in partially aligning microcrystals within the sample: they were placed in X-ray capillaries and pulled down by gently sucking their mother liquor, whereupon the needle-shaped crystals oriented along the capillary axis. Such samples sometimes, but only under perfect measuring conditions, produced pseudo fiber patterns which consisted of oriented arcs with average length of 60° (Fig. 1b). In many cases these arcs were composed of distinct spots that could be clearly resolved by eye. Since such patterns may arise from partial orientation of the nucleic acid component within fairly well packed particles, these patterns indicated reasonable internal order.

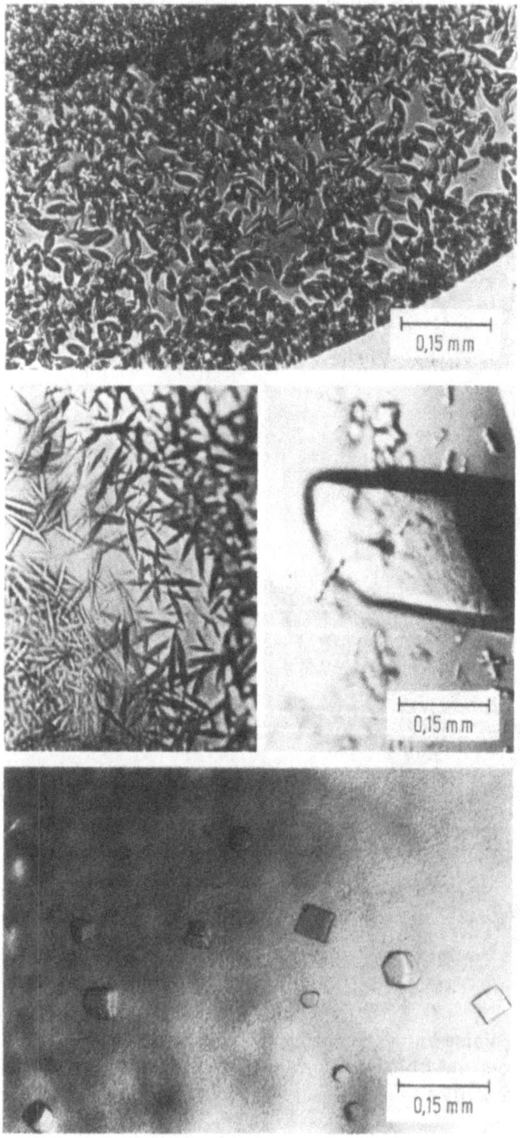

a

b

c

Fig. 2a–c. Different crystal forms of B. stearothermophilus 50S subunits (bar length 0.15 mm) **a** grown from 10% toluene, 0.5M NaCl at pH 6.2 (MES-buffer); **b** grown from 2.5% PEG 6000, 0.18M ammonium sulfate, 0.03M $MgCl_2$, and 0.09M KCl at pH 6.4 (MES-buffer) (left) and grown from 2.5% PEG 6000, 0.18M ammonium sulfate, 0.03M $MgCl_2$, and 0.54M KCl at pH = 6.6 (MES-buffer) (right); **c** grown from 2.5% PEG 6000, 0.015M ammonium sulfate, and 0.02M $MgCl_2$ at pH 6.4 (MES-buffer)

In favourable cases larger (though still tiny) crystals produced the first X-ray patterns on which distinct reflexions with periodic spacings could be detected (Fig. 1), thus encouraging further efforts.

Increasing the size of the crystals as well as their internal order by slowing down the crystallization process failed, probably because of deterioration of the particles before they could form aggregates; the latter seem to play an essential role in the nucleation process [16].

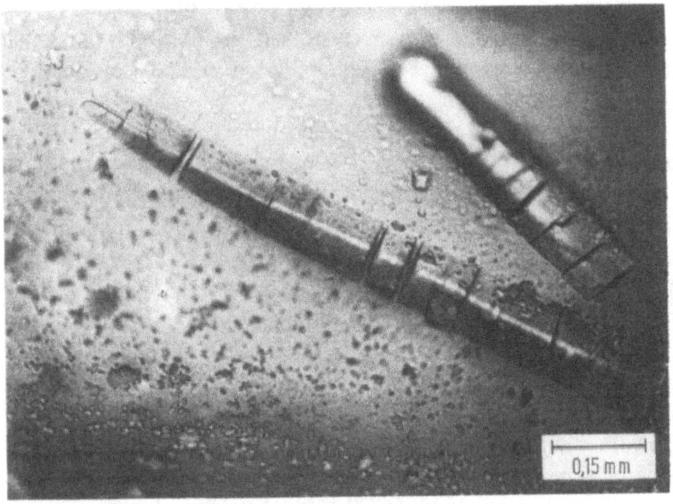

Fig. 3. Crystals of B. stearothermophilus 50S subunits grown in X-ray capillaries from methanol/ethylene glycol (at pH 8.4). Cracks develop within a few hours after sealing the capillaries [20] (bar length = 0.15 mm)

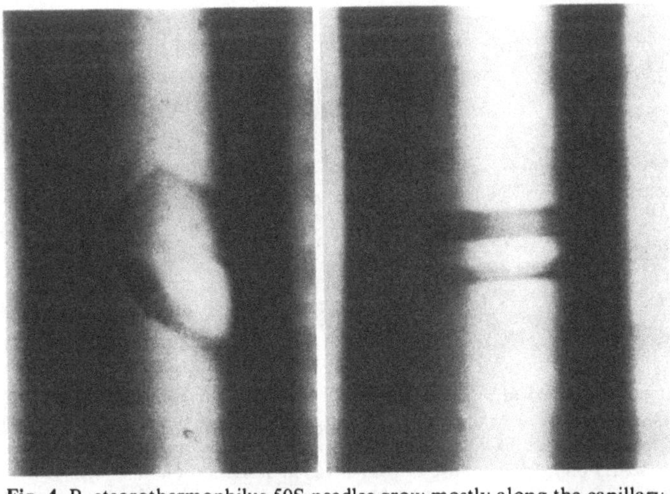

Fig. 4. B. stearothermophilus 50S needles grow mostly along the capillary axis but some of them also grow in different directions [20] (same conditions as in Fig. 3; capillary diameter = 0.7 mm)

A systematic exploration of parameters influencing crystal growth in ribosomal systems, and a fine control of the volume of the crystallization droplet by balancing water and alcohol diffusion through addition of salt to the reservoir [17)], eventually led to a number of different crystal forms for B. stearothermophilus 50S subunits (Figs. 2–4; cf. [18–22)]). Long pointed needles of up to 1.5 mm length grew directly in X-ray capillaries by vapor diffusion of alcohol mixtures. These crystals were stable for three to five months when kept in their natural growth medium at 4 to 7 °C.

Since precipitants used for growth were volatile alcohols (methanol, or methanol/ethylen glycol mixtures) any handling of the crystals was impossible. Even a minor intervention such as sealing the X-ray capillaries shortened the lifetime of crystals drastically to 6–8 hours (Fig. 3); crystals had to be irradiated immediately in their original growth solution.

Most of the crystals grew with their long axis parallel to the capillary axis, but a fair number of them grew in different directions (Fig. 4). Thus we were able to obtain diffraction patterns from all of the crystallographic zones without manipulating the crystals [20, 21)] (Fig. 5), despite the natural overcrowding of crystals which often imposes difficulties in finding solitary crystals in proper orientations (Fig. 5c).

All patterns showed a sharp decay in intensities at about 20 to 22 Å. Thus we were misled into believing that 20 Å was the actual resolution limit. Only much later, when crystals grew larger and measuring facilities were further improved, did we discover that the crystals are ordered internally to much higher level, close to that indicated by the "powder" diffraction.

Since we had to irradiate crystals within their growth solution in their original orientation, and at the same time to avoid other crystals in close proximity, we were strongly dependent not only on the intensity and the stability, but also on the size and shape of the incident beam. As a consequence, improvements in the quality of the data collection facilities were needed, alongside our efforts to improve the crystal quality.

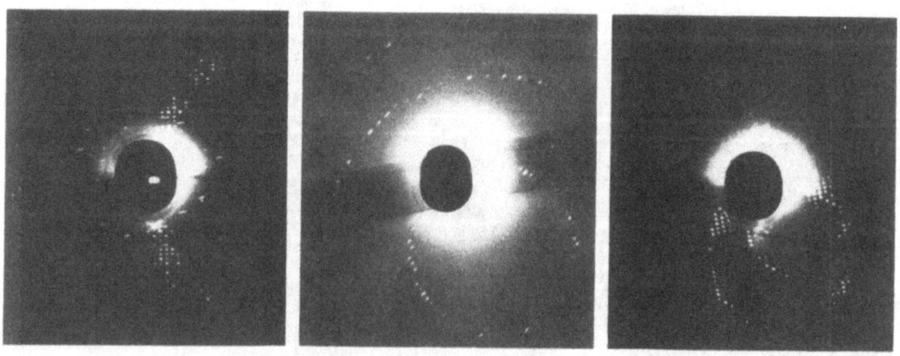

Fig. 5a–c. Diffraction patterns (obtained at 9.6/SRS/Daresbury and A1/CHESS/Cornell U.) from crystals like those in Fig. 4, exposed at −2 °C. **a** The twofold screw axis c = 905 Å can be seen on the edge of the 0k1-lune; rotation range 0.8°. Spurious streak-like reflexions result from neighbouring crystals in the capillary. **b** Still photograph close to the (tilted) hk0-orientation, a = 350 Å, b = 670 Å. **c** This picture demonstrates the difficulty in selecting a single crystal within the cross section of the X-ray beam

Among other developments, a new collimator system composed of two double slits was designed and built at EMBL/DESY [23]. Using the double focussing X-ray EMBL instruments equipped with this collimator system, we obtained sharp, resolvable patterns with enhanced signal to noise ratio, and we were able to determine cell dimensions [21].

2.2 Halobacterium Marismortui

Bearing in mind the technical difficulties arising from volatile organic solvents as precipitants, we also looked for ribosomes that are stable under high salt concentrations; they could perhaps be crystallized using "conventional" precipitants such as ammonium sulfate or other non-volatile agents. Thus the crystals could be handled with less difficulties and mounted in the conventional way in X-ray capillaries.

Ribosomes from halophilic bacteria meet this requirement; furthermore, the structure of a ribosome of an archaebacterium (Halobacterium marismortui) as compared to that of a eubacterium (Bacillus stearothermophilus) might prove a valuable extension of our studies.

The first microcrystals of H. marismortui 50S subunits were obtained at 4 °C from PEG in growth solutions that mimic the natural environment within these bacteria: Potassium, ammonium, magnesium and chloride ions were present in the crystallization solution at the minimum concentrations needed for reserving their activity [24].

Later on, advantage was taken of the delicate equilibrium of mono- and divalent ions needed for the growth of halobacteria. Crystals then grew at the minimum concentration of salts needed for storage without activity loss. They appeared as thin plates with dimensions of about 0.4 × 0.4 × 0.1 mm at 19 °C, but tended to form multilayer aggregates [25] (Fig. 6). Their X-ray diffraction patterns extended to about 13 Å resolution.

Fig. 6. Thin plates of H. marismortui 50S subunits grow spontaneously from PEG in the presence of salts (KCl) in hanging drops but tend to form multilayer aggregates [25] (bar length = 0.15 mm)

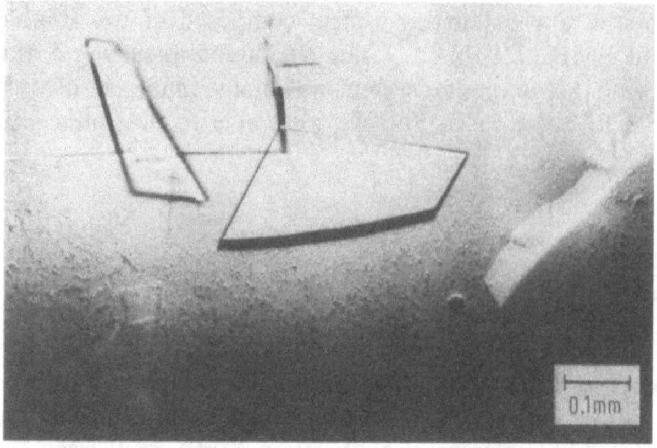

Fig. 7. Crystals of H. marismortui 50S subunits grown after seeding [27]. The face displays the monoclinic angle of 71° of the reduced cell (bar length = 0.1 mm)

It was found for spontaneous crystal growth of ribosomal particles that the lower the Mg^{2+} concentration, the thicker the crystals [26]. Consequently, crystals from 50S subunits from H. marismortui, grown spontaneously under the lowest Mg^{2+} concentration possible, were transferred as seed crystals to solutions with even a lower Mg^{2+} concentration. As a result, after about two weeks well ordered and relatively thick crystals of about 0.6 × 0.6 × 0.2 mm were formed (Fig. 7) that diffracted to about 6 Å [27].

3 Crystallographic Data Collection — Recent Developments

Only from 1985 onwards could we seriously consider crystallographic data collection. In fact, several full sets of oscillation films have since been exposed on the rotation camera; however, they have tended to become outdated in view of new developments in the crystallization techniques or by other experimental improvements.

3.1 Bacillus Stearothermophilus

It initially seemed most promising to use the abovementioned crystal form of 50S subunits of B. stearothermophilus grown directly in capillaries. These crystals may now reach a length of 2.0 mm and a cross-section of 0.4 mm, and are loosely packed in an orthorhombic unit cell of 350*670*910 Å [21]. Fresh crystals diffract to 10–13 Å resolution. However, the higher resolution reflexions decay within 5 to 10 minutes.

Hundreds of crystals were exposed in several synchrotron beam time periods, bearing in mind that there was little chance to solve the inherent problem of how to produce derivative crystals. It was a relief though, that 50S particles from a mutant of B. stearothermophilus lacking protein L11 could be crystallized isomorphously [28], which might serve as a low resolution deficiency derivative.

3.2 Halobacterium Marismortui

Seeded crystals grow in the orthorhombic space group $C222_1$ and diffract to a resolution of up to 6 Å (Fig. 8). They have relatively small, densely packed unit cells of 215*300*590 Å, in contrast to the "open" structure and the large unit cells of the

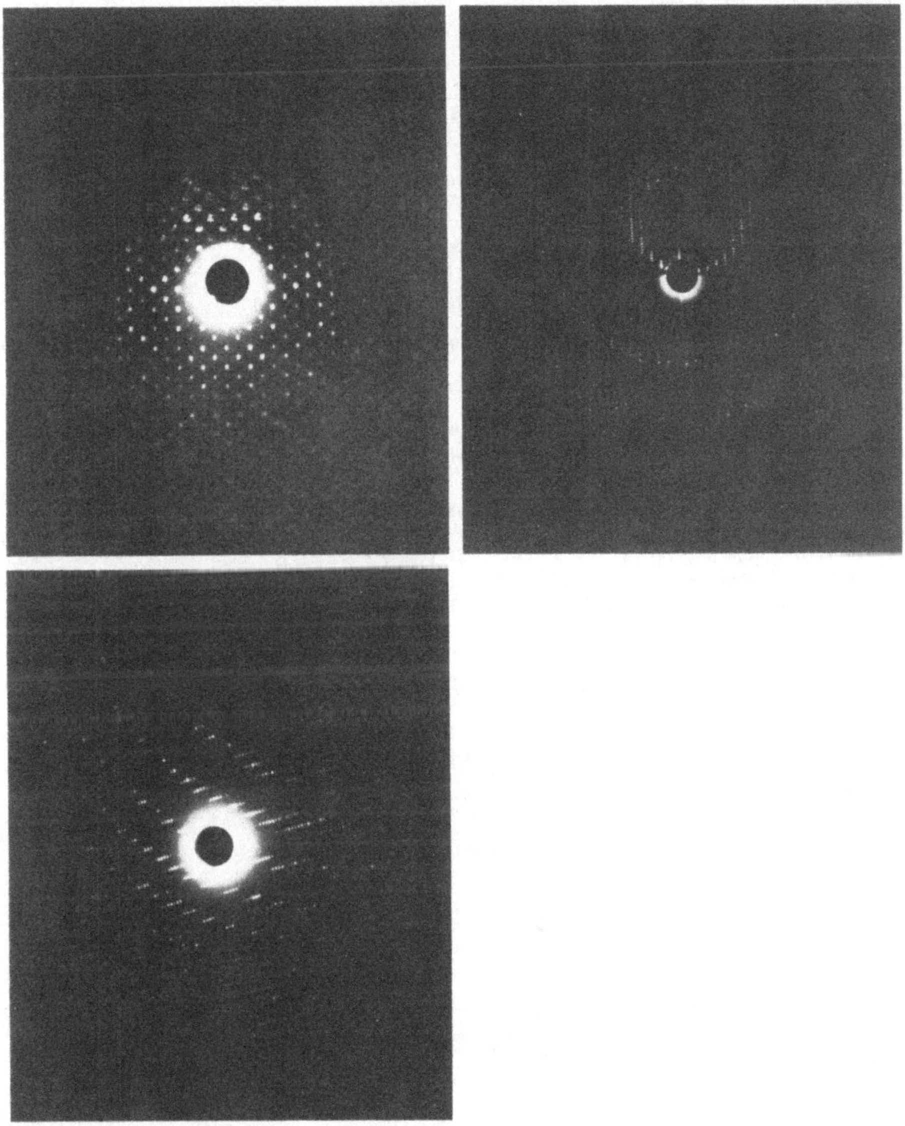

b

c

Fig. 8a–c. 1° rotation photographs of H. marismortui 50S crystals at 0 °C and at cryotemperature (obtained at X11/EMBL/DESY and at SSRL/Stanford U.) **a** The hk0-orientation of a nearly perfectly aligned (although split) crystal reveals the mirror symmetry of the C-centred lattice plane. The severe overlap problem in this orientation caused by the large mosaic spread is obvious from this picture. **b** The 0kl-orientation shows the extinctions of the twofold screw axis. **c** The best crystals have a Bragg resolution limit of about 6 Å, which decreases to about 9 Å in the course of a hundred exposures

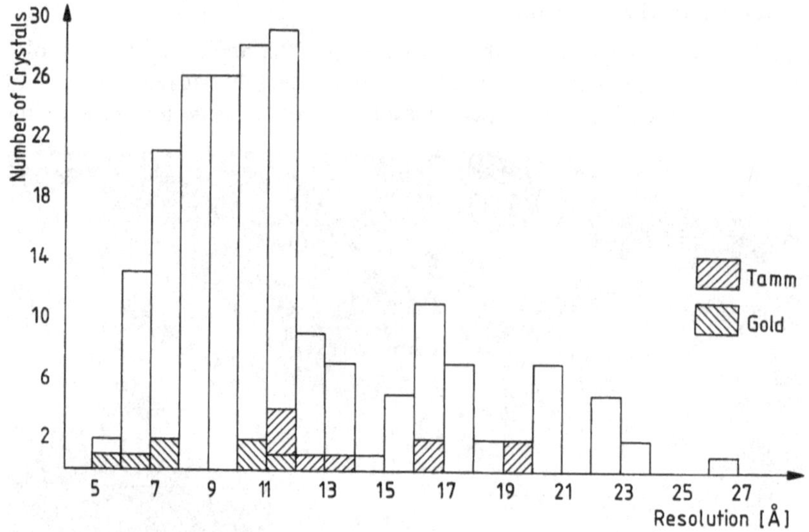

Fig. 9. Approximate Bragg resolution for the first exposure of each of about 200 crystals from H. marismortui 50S subunits that were investigated at X11, EMBL/DESY, Hamburg (FRG), in August 1986 at —4° to 19 °C. Shading indicates heavy-atom derivative test crystals (undecagold-cluster and tetrakis(acetoxymercuri)methane (TAMM); see paragraph 4: Phase Determination)

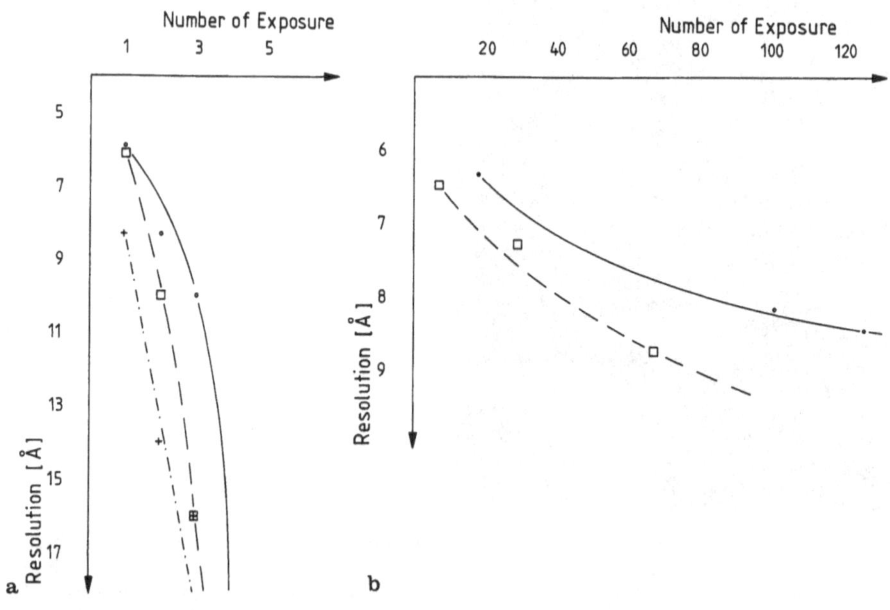

Fig. 10a and b. Decrease of Bragg resolution with time for crystals from H. marismortui 50S subunits when irradiated at —4° to 19 °C (left) and at cryotemperature (right). (Different symbols represent different crystals.)

crystals from 50S subunits of B. stearothermophilus. Up to 10 photographs can be taken from an individual crystal between −2 °C and 19 °C but the high resolution reflexions appear only on the first 1–3 X-ray photographs. Hence, over 260 crystals had to be irradiated in order to obtain a (supposedly) complete set of film data.

Crystals were aligned only visually to avoid the loss of precious reflexions in the higher resolution range in the course of setting photographs. None of the data films happened to show the orthorhombic symmetry, thus our preliminary report specified the approximate cell constants for the primitive monoclinic cell $P2_1$ [27]. Fig. 9 shows the statistical distribution of approximate Bragg resolution found for the first film from each crystal, Fig. 10a displays the quick loss in resolution as a function of the exposure time/film number. In order to average out the resolution decay during each exposure, the camera rotation axis was oscillated 10 times per photograph.

However, most crystals have a very large mosaic spread of up to 3° even in the first picture. This is indeed larger than the permissible rotation range at our current resolution. Hence there is a severe problem with overlap, in particular when the long 590 Å axis is in the direction of the X-ray beam. What is worse, for many crystals we are left with no fully recorded reflexions to scale the partial intensities ("post refinement", [29]). These problems, taken together with the short lifetime render even a joint refinement of all the three axes of individual crystals nearly impossible.

All this led to the question whether much lower temperatures might help to increase the lifetime of the crystals in the X-ray beam. We were introduced to cryotemperature crystallography, and learned how to embed the crystals in inert oil or similar materials before deep-freezing [30] instead of keeping them in capillaries. With this technique at least some crystals with Bragg resolution of about 6 Å could be transferred to liquid nitrogen temperature and would "live" in the synchrotron beam for many hours or even days (Fig. 10b). Figure 11 displays the initial Bragg resolution of those crystals that survived the procedure. As might be expected, the mosaic spread did not become smaller in the course of all this treatment, but it did not become much worse either.

In order to extract the crystals from their solution, tiny spatulas were constructed from very thin glass "saucers" glued to glass rods, which could be mounted on the

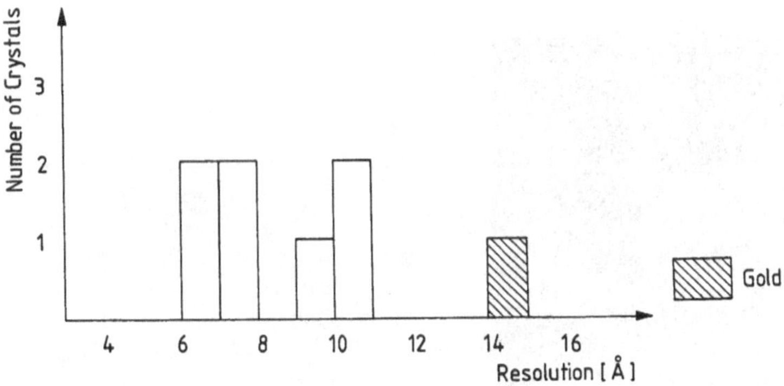

Fig 11. Approximate Bragg resolution for the first exposure of each of eight crystals (incl. one gold derivative) from H. marismortui 50S subunits that were investigated at SSRL beam-line 7-1 (Stanford University, Ca., USA) at cryotemperature

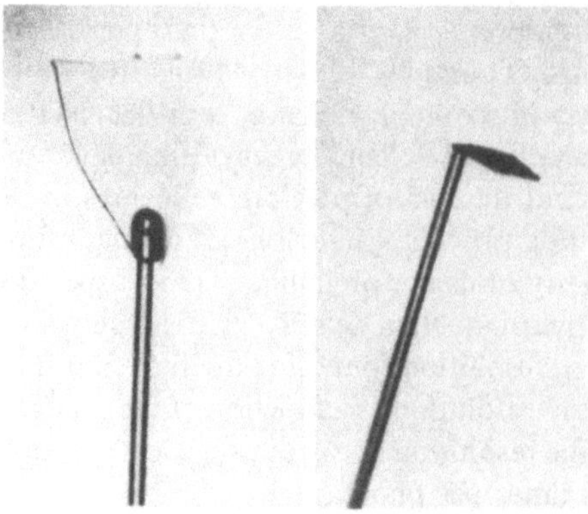

Fig. 12. Views of flat (left) and perpendicular (right) glass spatulas constructed for crystallographic studies of plate-like crystals from Halobacterium marismortui 50S subunits at cryotemperature. The long c-axis (i.e. the shortest crystal dimension) will coincide with the spindle axis (and overlap problems will be less severe) when perpendicular spatulas are used

goniometer head in a way familiar to small-molecule crystallographers. The first spatulas were flat, causing the same overlap problems as before. Now we have learned to glue the spatulas at different angles (Fig. 12) so as to preorient the crystal plates in directions that minimise the overlap problem.

There was some hope that the successes with cryotemperature methods would make progress less dependent on a high intensity synchrotron beam and that at least preliminary experiments such as the search for isomorphous derivatives would become feasible in our own laboratory. However, experience in the past year has taught us that neither a weaker synchrotron beam nor a rotating anode ever yields a diffraction pattern comparable in Bragg resolution to Figure 8c.

4 Phase Determination

The large size of ribosomal particles is an obstacle for crystallographic studies, but permits direct investigation by electron microscopy. A model (Fig. 13) obtained by three-dimensional image reconstruction of two-dimensional sheets (e.g. [31]) may be used for gradual phasing of low resolution crystallographic data.

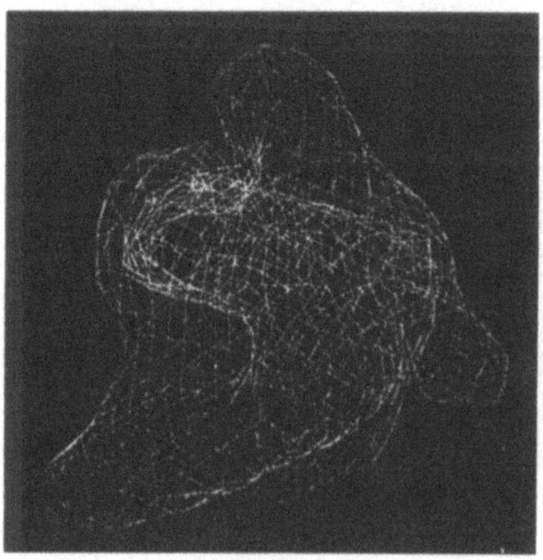

Fig 13. Computer graphics display of the model of the B. stearothermophilus 50S subunit reconstructed from 2-dimensional sheets at 30 Å resolution

Heavy-atom derivation of an object as large as a ribosomal particle requires the use of extremely dense and ultraheavy compounds. Examples of such compounds are a) tetrakis(acetoxy-mercuri)methane (TAMM) which was the key heavy atom derivative in the structure determination of nucleosomes [32] and the membrane reaction center [33], and b) an undecagold cluster in which the gold core has a diameter of 8.2 Å (Fig. 14 and in [34] and [35]). Several variations of this cluster, modified with different ligands, have been prepared [36]. The cluster compounds, in which all the moieties R (Fig. 14) are amine or alcohol, are soluble in the crystallization solution of 50S subunits from H. marismortui. Thus, they could be used for soaking. Crystallographic data (to 18 Å resolution) show isomorphous unit cell constants with observable differences in the intensities (Fig. 15).

Because surfaces of ribosomal particles have a variety of potential binding sites for such clusters, attempts are in progress to bind heavy-atoms covalently to a few specific sites on the ribosomal particles prior to crystallization. This may be achieved either by direct interaction of a heavy-atom cluster with chemically active groups such as -SH or the ends of rRNA [37] on the intact particles or by covalent attachment of a cluster to natural or tailor-made carriers that bind specifically to ribosomes.

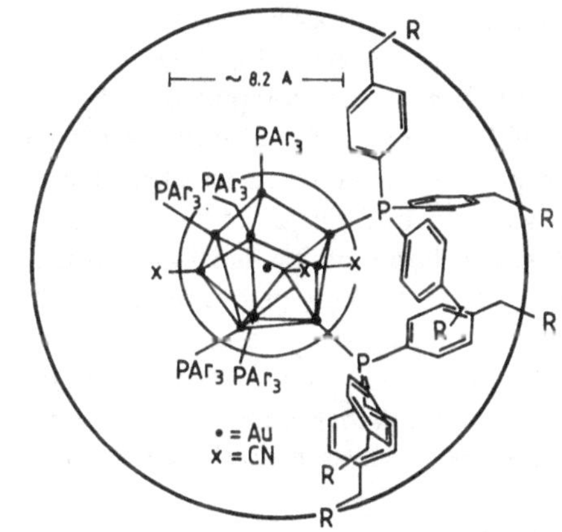

Fig. 14. Semi-schematic presentation of the undecagold cluster depicting the gold core of 8.2 Å diameter, and the arrangement of ligands around it (after [43] and literature cited therein)

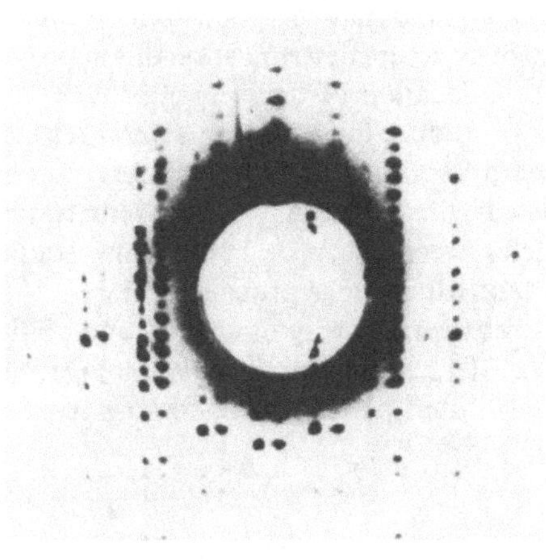

Fig. 15. Superposition of the central reflexion patterns (obtained at SSRL/Stanford U.) of a native and an undecagold derivative crystal (rotation photographs of nearly identical orientations, slightly shifted against each other to facilitate comparison of intensities)

For direct binding to the surfaces of the ribosomes, the following approaches were used: Firstly, free sulfhydryls on the surface of the 50S subunit were located by reacting with radioactive N-ethylmaleimide. The labeled proteins were identified by locating radioactivity in two-dimensional gels. For 50S subunits from B. stearothermophilus, the two proteins BL11 and BL13 appear to bind N-ethylmaleimide. For H. marismortui a significant portion of the radioactivity was associated with a single protein, tentatively named HL11.

Secondly, the gold cluster described above was prepared in such a way that it could be bound to accessible -SH groups. Since this cluster is rather bulky, its accessibility was increased by the addition of spacers of various lengths to the cluster and to the free -SH groups on the ribosomal particles.

Radioactive labeling of this cluster and neutron activation analysis of the gold enabled us to determine the extent of binding of the cluster to the particles. The results of both analytical methods show that a spacer of minimum length of about 10 Å between the -SH group of a ribosomal protein and the N-atom on the cluster is needed for significant binding. Preliminary experiments indicate that the products of the derivatization reaction with 50S particles can be crystallized.

As mentioned above, such clusters may also be bound to biochemical carriers. Examples of these are antibiotics [38)], DNA oligomers complementary to exposed single-stranded rRNA regions [39)] and Fab fragments of antibodies specific to ribosomal proteins. Most of the interactions of these compounds have been characterized biochemically; the crystallographic location of the heavy-atom compounds will thus not only be used for phase determination but will also reveal the specific functional sites on the ribosome. Alternatively, such clusters may be attached to selected sites on isolated ribosomal components; the latter will subsequently be incorporated into particles that lack them. Thus, a mutant of B. stearothermophilus that lacks protein BL11 was obtained by growing cells in the presence of thiostrepton at 60 °C [40)]. The 50S mutated ribosomal subunits yield sheets and crystals, isomorphous with those of 50S from the wild-type [28)]. This indicates that BL11 is not involved in crystal forces in the native crystals. Furthermore, protein BL11 has only one sulfhydryl group, and binding of N-ethylmaleimide to it does not reduce the biological activity and crystallizability of the particles.

Since protein BL11 is nearly globular [41)] its location may be determined in a Patterson map with coefficients of $[F(wild)-F(mutant)]^2$ and may serve, by itself, as a giant heavy-atom derivative. At preliminary stages of structure determination this approach may provide phase information and reveal the location of the lacking protein.

Correspondingly, we have also adopted a procedure for removing several selected proteins from the ribosomal subunits of B. stearothermophilus [42)]. The deleted proteins were, in turn, incorporated into the depleted core particles and the activity and crystallizability of the reconstituted particles were checked. Preliminary studies reconfirm the results obtained with the mutant which lacks protein BL11.

Furthermore, there is a good correlation between recovery of activity and ability to crystallize. Thus, particles lacking protein BL12, which are inactive biologically, could not be crystallized, whereas the reconstituted particles produce crystals isomorphous with the native form.

5 Summary

We have shown that out of fifteen forms of three-dimensional crystals from ribosomal particles, grown so far in our laboratory, some appear suitable for crystallographic data collection when using synchrotron radiation at temperatures between 19 °C and −180 °C: 50S subunits from H. marismortui., and from B. stearothermophilus, including the -BL11 mutant, and the new crystal forms from B. stearothermophilus 50S and Thermus thermophilus 30S subunits which have only recently been grown in non-volatile precipitants [22]. We also plan to continue research on biochemically modified particles, such as 50S with one tRNA and its nascent polypeptide chain (which have already been crystallized).

All this should eventually lead to a three-dimensional model which, (if not at the atomic level), should show molecular details that may assist in the understanding of the interaction of the ribosome with the variety of other components which cooperate in the biosynthetic process.

6 Acknowledgements

We wish to thank Dr. H. D. Bartunik, K. Wilson, J. Helliwell, M. Papiz, K. Moffat, W. Schildkamp, P. Phizackerley, and E. Merrit, and their respective staff for their support on the synchrotron radiation facilities EMBL/DESY, SRS, CHESS and SSRL.

We are grateful to Drs. H. Hope and C. Kratky who have introduced us to and extensively collaborated on using cryotemperatures, to Drs. F. Frolow and M. A. Saper for their efforts in data collection, to our students K. v. Boehlen and K. Stegen, and to H. Danz, H. S. Gewitz, C. Glotz, Y. Halfon, G. Idan, I. Makowski, J. Müssig, J. Piefke, B. Romberg, and P. Webster for skillful technical assistance.

This work was supported by BMFT (05 180 MP BO), NIH (GM 34360) and Minerva research grants.

7 References

1. Chambliss, G., Craven, G. R., Davies, J., Davies, K., Kahan, L., Nomura, M. (eds): Ribosomes: Structure, Function, and Genetics. Baltimore, Univers. Park Press 1980
2. Wittmann, H. G.: Ann. Rev. Biochem. *51*, 155 (1982)
3. Wittmann, H. G.: ibid. *52*, 35 (1983)
4. Liljas, A.: Progr. Biophys. Mol. Biol. *40*, 161 (1982)
5. Hardesty, B., Kramer, G. (Eds.): Structure, Function, and Genetics of Ribosomes, Springer-Verlag, Heidelberg and New York 1986
6. Byers, B.: J. Mol. Biol. *26*, 155 (1967)
7. Unwin, P. N. T.: Nature *269*, 118 (1977)
8. Milligan, R. A., Unwin, P. N. T.: ibid *319*, 693 (1986)
9. Yonath, A. E., Müssig, J., Tesche, B., Lorenz, S., Erdmann, V. A., Wittmann, H. G.: Biochem. Intern. *1*, 428 (1980)
10. Yonath, A., Wittmann, H. G.: Biophys. Chem. *29*, 17–29 (1988)
11. Trakhanov, S. D., Yusupov, M. M., Agalarov, S. C., Garber, M. B., Ryazantsev, S. N., Tischenko, S. V., Shirokov, V. A.: FEBS Letters *220*, 319 (1987)

12. Wittmann, H. G., Müssig, J., Piefke, J., Gewitz, H. S., Rheinberger, H. J., Yonath, A.: ibid. *146*, 217 (1982)
13. Zubay, G., Wilkins, M. H. F.: J. Mol. Biol. *2*, 105 (1960)
14. Klug, A., Holmes, K. C., Finch, J. T.: ibid. *3*, 87 (1961)
15. Langridge, R., Holmes, K. C.: ibid. *5*, 611 (1962)
16. Yonath, A., Khavitch, G., Tesche, B., Müssig, J., Lorenz, S., Erdmann, V. A., Wittmann, H. G.: Biochem. Intern. *5*, 629 (1982)
17. Yonath, A., Müssig, J., Wittmann, H. G.: J. Cell. Biochem. *19*, 145 (1982)
18. Yonath, A., Tesche, B., Lorenz, S., Müssig, J., Erdmann, V. A., Wittmann, H. G.: FEBS Letters *154*, 15 (1983)
19. Yonath, A., Piefke, J., Müssig, J., Gewitz, H. S., Wittmann, H. G.: ibid. *163*, 69 (1983)
20. Yonath, A., Bartunik, H. D., Bartels, K. S., Wittmann, H. G.: J. Mol. Biol. *177*, 201 (1984)
21. Yonath, A., Saper, M. A., Makowski, I., Müssig, J., Piefke, J., Bartunik, H. D., Bartels, K. S., Wittmann, H. G.: ibid. *187*, 633 (1986)
22. Glotz, C., Müssig, J., Gewitz, H. S., Makowski, I., Arad, T., Yonath, A., Wittmann, H. G.: Biochem. Intern. *15*, 953 (1987)
23. Bartunik, H. D., Gehrmann, T., Robrahn, B.: J. Appl. Crystallogr. *17*, 120 (1984)
24. Shevack, A., Gewitz, H. S., Hennemann, B., Yonath, A., Wittmann, H. G.: FEBS Letters *184*, 68 (1985)
25. Shoham, M., Müssig, J., Shevack, A., Arad, T., Wittmann, H. G., Yonath, A.: ibid. *208*, 321 (1986)
26. Arad, T., Leonard, K. R., Wittmann, H. G., Yonath, A.: The EMBO Journal *3*, 127 (1984)
27. Makowski, I., Frolow, F., Saper, M. A., Shoham, M., Wittmann, H. G., Yonath, A.: J. Mol. Biol. *193*, 819 (1987)
28. Yonath, A., Saper, M. A., Frolow, F., Makowski, I., Wittmann, H. G.: ibid. *192*, 161 (1986)
29. Rossmann, M. G., Leslie, A. G. W., Abdel-Meguid, S. S., Tsukihara, T.: J. Appl. Crystallogr. *12*, 570 (1979)
30. Hope, H.: Acta Crystallogr. *B44*, 22–26 (1988)
31. Yonath, A., Leonard, K. R., Wittmann, H. G.: Science *236*, 813 (1987)
32. Richmond, T., Finch, J. T., Rushton, B., Rhodes, D., Klug, A.: Nature *311*, 533 (1984)
33. Deisenhofer, J., Epp, O., Mikki, K., Huber, R., Michel, H.: J. Mol. Biol. *180*, 385 (1984)
34. Bellon, P., Manassero, P. M., Sansoni, M.: J. Chem. Soc. Dalton Trans., 1481 (1972)
35. Wall, J. S., Hainfeld, J. F., Barlett, P. A., Singer, S. J.: Ultramicroscopy *8*, 397 (1982)
36. Weinstein, S., Jahn, W.: private communication
37. Odom Jr., O. W., Robbins, D. R., Lynch, J., Dottavio-Martin, D., Kramer, G., Hardesty, B.: Biochem. *19*, 5941 (1980)
38. Nierhaus, K. H., Wittmann, H. G.: Naturwissenschaften *67*, 234 (1980)
39. Hill, W. E., Trappich, B. E., Tassanakajohn, B.: Probing Ribosomal Structure and Function, in: Structure, Function and Genetics of Ribosomes (ed.) Hardesty, B. and Kramer, G., p. 233, Springer-Verlag, Heidelberg and New York 1986
40. Schnier, J., Gewitz, H. S., Leighton, B.: to be published
41. Giri, L., Hill, W. E., Wittmann, H. G., Wittmann-Liebold, B.: Adv. Prot. Chem. *36*, 1 (1984)
42. Gewitz, H. S., Glotz, C., Goischke, P., Romberg, B., Müssig, J., Yonath, A., Wittmann, H. G.: Biochem. Intern. *15*, 887 (1987)
43. Reardon, J. E., Frey, P. A.: Biochem. *23*, 3849 (1984)

Application of EXAFS to Biochemical Systems

S. Samar Hasnain

Synchrotron Radiation Source, SERC Daresbury Laboratory, Warrington WA4 4AD, Cheshire, U.K.

Table of Contents

1 Introduction

X-ray absorption spectroscopy is an exciting new tool, ideally suited to probing the immediate environment of a specific atom type in a physical, chemical or biological system. The advent of synchrotron radiation has transformed this technique from a topic of relatively minor interest to one of major scientific importance and activity [1-9]. A major attraction of the technique is the possibility it provides of probing a reaction centre in a wide range of materials ranging from an industrial catalyst to an enzyme; the technique is not limited by the physical state of the sample. In this review, suitability of this technique for biochemical systems is discussed.

Figure 1 describes the basis of the technique schematically. The atom of interest is selectively excited to generate the electron, which when reflected back from its surroundings, interferes with itself to give oscillatory behaviour in the x-ray absorption spectrum. The finite electron mean free path limits the probing sphere to less than 10 Å. This local nature of the process gives the technique one of its major strengths, in that the structure around a specific atom type can be determined very accurately. For example, the metal-ligand distances can be determined to an accuracy of better than 0.02 Å. The specificity of information thus obtained, along with accurate chemical information, are invaluable in determining structural changes upon e.g. reduction, pH changes or binding of substrate analogue and inhibitors.

Since the first experiment on an intense synchrotron radiation source a little over ten years ago [10] x-ray absorption spectroscopy (XAS) has become a routine technique for investigating the environment of metal ions in chemical and biochemical systems, and numerous studies have been reported in the literature. It is not the purpose of the present article to deal with these studies individually and the reader is referred to several reviews which have been published [1-9]. A few examples will be given to illustrate the potential of the technique, and to give some indication of the type of experiments which may be considered for studies in the future.

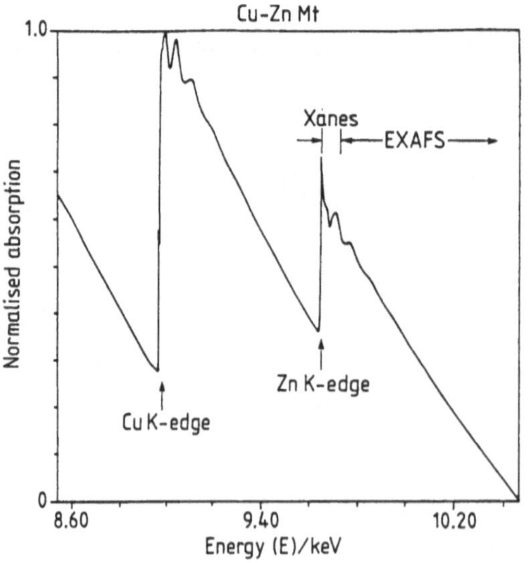

Fig. 1. X-ray absorption spectrum (XAS) of Cu—Zn metallothionein at the Cu and Zn K-edges. The structure near the edge, referred to as XANES is dominated by multiple scattering events while the extended structure, referred to as EXAFS, at photoelectron energies greater than 30–50 eV is primarily due to single scattering events

2 Experimental Requirements

The EXAFS oscillations are superimposed on a smooth but much larger background absorption resulting from the electronic excitation of the metal atom. Thus, high quality EXAFS data cannot be obtained unless most stringent experimental conditions are met in terms of x-ray source, optics and detectors. In this section, some of these aspects are discussed.

2.1 X-ray Source and Synchrotron Radiation

Although x-ray absorption spectra have been measured for nearly sixty years [11] using conventional x-ray sources, synchrotron radiation has provided the major breakthrough. Synchrotron radiation is emitted when a beam of electrons or positrons is constrained to travel on a circular trajectory under the influence of a magnetic field. The radiation is emitted in a continuous wavelength spectrum around a characteristic wavelength λ_c which is inversely proportional to the square of the energy of the electron beam E and to the field of the magnet B,

$$\lambda_c(\text{Å}) = \frac{18.65}{E^2(\text{GeV})\, B(\text{T})} \tag{1}$$

Since the first experiment on a synchrotron radiation beam line in 1975 [10], synchrotron radiation sources have seen major developments with the design of dedicated synchrotron radiation sources, such as the SRS in the United Kingdom, Photon Factory in Japan and NSLS in the United States. Some of the high energy physics storage rings built for colliding beams research e.g. Doris (at Hamburg) and Spear (at Stanford) are devoted for a large percentage of their operational time to synchrotron radiation research.

Figure 2 shows the experimental arrangement involved with the production of synchrotron radiation at the U.K.'s Synchrotron Radiation Source at Daresbury Laboratory [12]. The principal components are similar to those used at the other synchrotron radiation facilities. Electrons obtained from an electron gun are accelerated in a linear accelerator and in a 'booster' synchrotron prior to injection into the main electron storage ring. The electrons are injected into the storage ring at 605 MeV and then slowly ramped to a working energy of 2.0 GeV with a lifetime of up to 16 hours. The electrons circulate in discrete 'bunches' at a frequency of 3 MHz with a maximum current of 300 mA involving 160 circulating bunches. Thus bursts of synchrotron radiation pulses (170 ps) are observed in the laboratory frame every 2 ns. It is also possible to operate the storage ring in a 'single bunch' mode and in this case the synchrotron radiation pulses are separated by the full circulating time of the electron bunch, 320 ns; thus providing a very useful time structure for the time-resolved studies. In addition to the time structure, there are a number of properties wich make synchrotron radiation a unique source of electromagnetic radiation. For x-ray spectroscopy the most relevant properties are:

(i) that it is continuous in the whole of the x-ray region, $\geqq 0.3\lambda_c$; thus it is possible to cover the K- and/or L-edges of all elements in the periodic table;

(ii) very high intensity, $>10^{11}$ photons/sec/eV, coupled with extreme source stability;

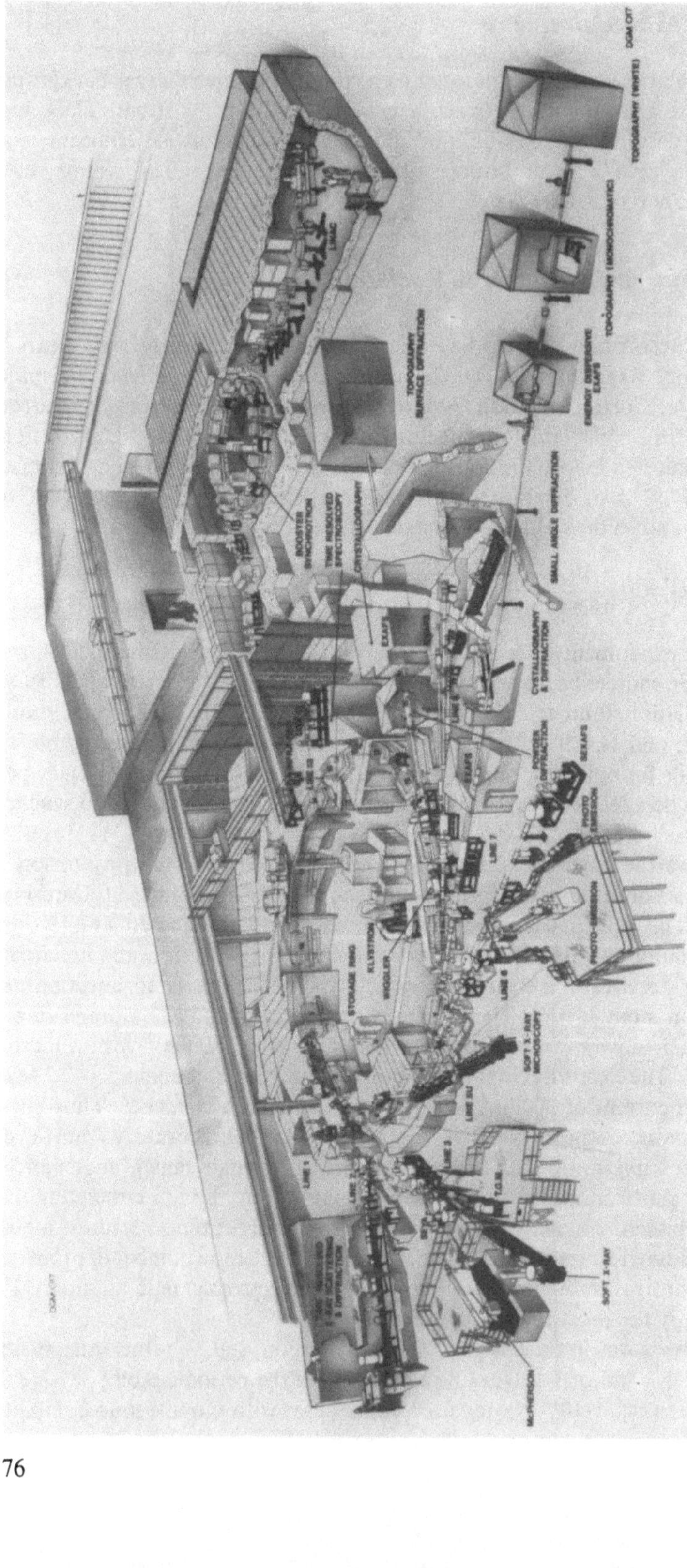

Fig. 2. Layout of the Synchrotron Radiation Source (SRS) at the Daresbury Laboratory, U.K.

thus it is possible to obtain high quality data of a protein, for example, in a reasonably short time (6–12 hours);

(iii) high vertical collimation due to the relativistic speed of the electrons; thus enabling the installation of focussing optics at some distance (necessary for radiation shielding) away from the source; and

(iv) high degree of transverse polarisation in the horizontal plane.

2.2 An Optimum XAS Instrument

The spectral purity, high brightness and extreme stability of the monochromatic beam are some of the most important requirements for an optimum XAS instrument. These are particularly crucial for biochemical systems such as metalloproteins as these are normally available in small quantities and have low metal concentration. In addition, there is some evidence that radiation damage of proteins is time dependent rather than radiation dose dependent; thus providing further incentive for optimising instrumentation such that maximum signal-to-noise can be obtained in the shortest measuring time. Extensive effort has been made at various synchrotron radiation centres in improving the performance of these instruments. In particular, major advances have been made in the use of focussing crystal optics [13–15]. In this section, I will attempt to report the advances briefly with the aim of outlining the design of an optimum XAS instrument for biological specimens.

X-rays from a storage ring are emitted in a narrow vertical cone, 0.2 mrad at 1 Å (40 arc sec) over the whole horizontal extent of the bending magnet or wavelength shifter. Both of these features have been appreciated for some time and at most synchrotron radiation centres XAS instruments have been designed to take advantage of these features. Thus, vertically dispersing optical arrangements have been implemented for XAS, where the diffraction widths of perfect crystals such as Si(111) and Si(220) match reasonably well with the vertical divergence of the x-rays from the storage rings. Figure 3 shows a typical layout of the experimental arrangement where a long ($\simeq$ 60 cm–75 cm) toroidal mirror is used to collect a significant horizontal aperture (2 to 3 mrad) of the radiation, which focusses it into a small spot at the

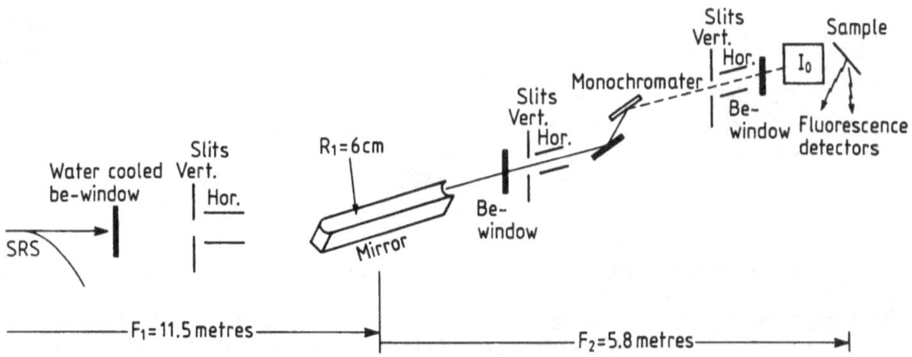

Fig. 3. A typical XAS instrument on a Synchrotron Radiation Source

sample. In some cases, the mirror is placed after the monochromator in order to reduce fluctuations in the output beam due to temperature variations with beam decay. The monochromator is typically a two-crystal arrangement with varying degrees of sophistication [13-23]. In most cases, attempts are being made to compensate for the beam heating effects by clamping the first crystal onto a cooled copper block with or without a liquid metal (gallium) interface. Pettifer et al. [23] have chosen to heat the first crystal so as to maintain a constant temperature. Most recent monochromators have also been designed with harmonic rejection and fixed exit beam capabilities. The former is achieved by tilting the diffracting planes of one crystal with respect to the other, while the latter is achieved primarily by translating the second or both crystals. Recently at Chess [16] and the Photon Factory [13] a sagittally bent crystal monochromator has been used successfully. In this case the second flat crystal is replaced by a sagittally bent crystal to provide the focussing. The focussing is optimised at an absorption edge and a small drift to the focus is accepted as the wavelength is scanned. Although the capability of rejecting higher harmonics by tilting the crystal has been built in, a small flat mirror is used at the Photon Factory to reject harmonics in order to keep the operation of the monochromator simple. This type of monochromator is still in its development stage and a number of improvements can be expected in the coming decade, including dynamic focussing, harmonic rejection, etc.

In most cases, a slit has been used to limit the vertical acceptance of the monochromator and/or mirror, in order to obtain the energy resolution which is required for XAS. Thus, only a fraction of the available beam intensity is utilised. Hoek et al. [14] have recently designed a novel two crystal slitless monochromator, where the crystals are bent and the radius is matched to the source divergence; thus, high energy resolution is obtained without limiting the acceptance angle. Hoek et al. have also tackled the crystal heating problem effectively by cooling the first crystals directly. In this case the holes are drilled in the crystal and the refrigerated water is passed through the crystal. In addition, harmonic rejection is achieved by tilting the second crystal and a fixed exit beam is provided by translating the second crystal onto an angled track. Focussing is provided by a toroidal mirror which is located after the monochromator, Fig. 4. For the high energy range where the mirror reflectivity is poor and the size of mirror required is rather long, a crystal pair with a second crystal

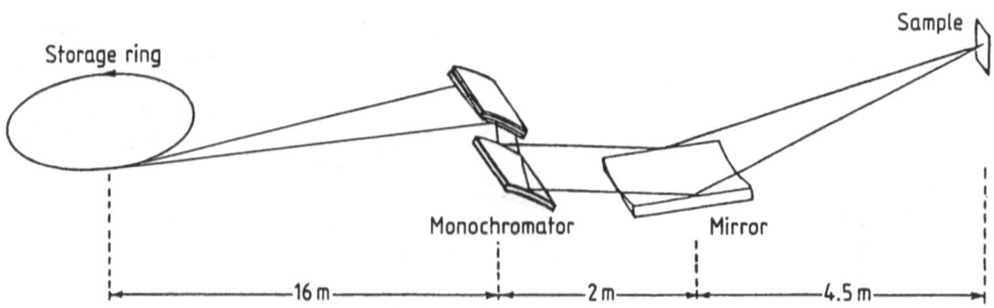

Fig. 4. Experimental arrangement on the XAS instrument 8.1 at the SRS, where a slitless monochromator is used and the crystals are mechanically bent to account for the vertical divergence (0.2 mrad) of the x-rays from the source. A toroidal mirror is placed after the monochromator, which provides the focussing

bent in both directions provides the focussing. This design, like the sagittal focussing monochromator, can be expected to improve further in the coming years, particularly with respect to achieving dynamical focussing and coping with the high power load expected from future sources such as the next generation of x-ray sources being planned at ESRF and Argonne. One irritating aspect of all XAS monochromators has been the presence of crystal glitches or Laue spots and their non-normalisation behaviour for samples with low metal concentrations [15]. In protein XAS this quite often limits the data range. We (Hasnain, Morrell, Dobson and Hart, unpublished work) have shown that the non-normalisation of crystal glitches primarily results from misalignment of the crystals with respect to each other and to the optic axis. Future designs will need to incorporate independent adjustment of crystals for accurate alignment.

In addition to an intense source and a well matched monochromator/mirror system, detectors require optimisation. For protein XAS, it is now well established that fluorescence detection is the preferred mode of detection [24–28]. Multi-detector systems using scintillation counters have been set up at most storage rings in order to improve the S/N for a given incident flux. These detection systems along with the focussing monochromator arrangements described above have provided good quality data for 1–2 mM protein solutions for 3d metals and about 0.5 mM protein solutions for metals such as molybdenum. This type of detection system relies on metal filters (made from (Z-1) or (Z-2) material) to reduce the x-ray scattering in preference to the x-ray fluorescence. Further improvements can be made if energy dispersive detectors are used. Cramer and George (private communication) have recently used a thirteen-element germanium detector system and have obtained high quality XAS data for 0.5 mM Mn in photosystem, thus providing an improvement of nearly a factor of four over the conventional scintillation detection system.

In summary, an optimum XAS instrument for biological specimens should include a slitless, fixed exit beam position, effectively cooled focussing monochromator along with a multi-element high count rate energy dispersive detector. We can expect such an XAS instrument on most of the synchrotron radiation sources in the next few years. Further developments can be expected with the realisation of next generation sources such as the 6 GeV storage ring at ESRF, Grenoble and the proposed 7 GeV storage ring at Chicago.

3 XAS Theory and Data Analysis

The interpretation of the EXAFS in terms of the local chemistry of a compound was shown to be possible by Lytle, Sayers and Stern in the early 1970's [29–31]. They formulated a single scattering short-range order theory which gave excellent agreement with the experimental data beyond 50 eV above the absorption threshold. They showed that the frequency and the amplitude of the EXAFS could be related to the interatomic distances and the coordination number around the photoexcited atom. They also pointed out that the experimental EXAFS data can be Fourier transformed to give a radial structure function containing direct information on the bond lengths, number of atoms and widths of coordination sphere around the excited atom.

Full derivations of the theory were presented by Lee and Pendry [32] and Ashley and Doniach [33] in 1975. They showed that a complete quantitative description of the EXAFS process was possible and that accurate bond lengths and coordination numbers could be extracted from the analysis of EXAFS data. Lee and Pendry also showed that at high photoelectron energies, the curvature of the electron wave can be neglected and thus the theory can be greatly simplified into what has become known as the plane-wave approximation. This approximation results in an expression equivalent to that derived by Stern [31] semi-empirically:

$$\chi(K) = \frac{\Delta\mu}{\mu_0} = -\sum_j \frac{N_j}{KR_j^2} |f_j(\pi)| \sin(2KR_j + 2\delta_l + \varphi_j)$$
$$\times \exp(-2\sigma_j^2 k^2) \exp(-2R_j/\lambda) \qquad (3.1)$$

The structural basis of EXAFS can be clearly appreciated from this expression, in that $\chi(K)$ is dependent on the number (N_j), location (R_j) and the type of scattering atoms, $|f_j(\pi)|$. The backscattering amplitude $|f_j(\pi)|$ has characteristic energy dependence. $2\delta_l$ is a phase shift due to the potential of the emitting atom which an electron experiences on leaving and re-entering the excited atom while ψ_j is the phase of the backscattering factor. A knowledge of the phase shift ($2\delta_l + \psi_j$) is required in the quantitative analysis of EXAFS data in order to obtain the correct interatomic distances, R_j. A Fourier transform without allowing for the phase shift correction results in an error of up to 20% in the bond distance. Two different approaches have been used for obtaining phase shifts. These rely either on the transfer of phase shifts derived from the EXAFS analysis of compounds of known structure (the empirical approach) or on the calculation of phase shifts (the ab initio method). The latter has the advantage that simple well-characterised chemical systems are not essential and that phase shifts can be calculated over the whole data range and thus use of the curved wave method of data analysis can be made. The phase shifts are calculated by constructing an atomic potential of the chemical cluster under study. The solution of the Schrödinger equation for an electron passing through this potential is compared with that of the free electron and the difference between them gives the phase shift for the electron wave of particular energy and angular momenta. The phase shifts are calculated for angular momenta 0 to 13. The detailed procedure for the calculation of phase shifts has been discussed elsewhere [34,35]. In our experience, the curved wave approach with ab initio phase shifts offers one of the most reliable methods for the extraction of structural data for metalloproteins if the phase shifts are adequately tested on related chemical systems. Gurman, Binsted and Ross [36] have simplified this exact curved wave theory by averaging the terms which describe the interaction between the polarisation vector of the radiation ($\vec{E}$) and the electron scattering vectors from the metal to the backscattering atom ($\vec{R}_j$). This simplification does not comprise the exact nature of the theory for polycrystalline or amorphous samples and is well suited for studies of metal ions in a protein solution. This simplification has resulted in a major improvement in the analysis program EXCURVE.

Equation (3.1) also assumes that the photoelectron is scattered off only one neighbouring atom i.e. EXAFS results from only a single scattering event. This is true for most cases particularly at relatively high energies (200 eV above the edge). However,

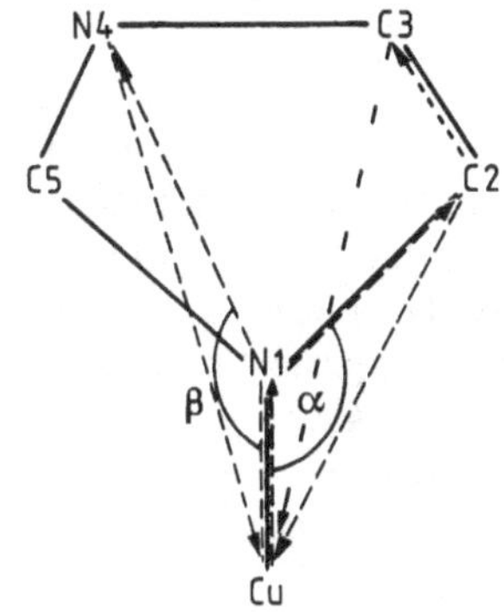

Fig. 5. Single and multiple scattering pathways which make a significant contribution to the EXAFS data of an imidazole-containing compound

in cases where three or more atoms are arranged collinearly or nearly so, then strong multiple scattering of the electron can take place [37, 38]. This was realised by Lee and Pendry [32] who performed detailed multiple scattering calculations and noted that due to the collinear arrangement of the first and fourth shells of copper in the metallic Cu, strong multiple scattering takes place. Several other examples of amplitude enhancement and phase modification due to multiple scattering have been noted including the enhanced oxygen backscattering in $Mo(CO)_6$ [39] and the outer shell carbon and nitrogen atoms of imidazole containing complexes [40]. Recently the single-scattering curved wave EXAFS function has been modified to include the double- and triple-scattering contributions [37]. In the curved wave theory, which works quite well for the low photoelectron energies, this is an important correction as the forward scattering from an intervening atom can be very strong over a significant angular range (0–70°). Figure 5 shows the multiple scattering pathways for an imidazole group. At higher photoelectron energies the forward scattering is increasingly confined to a narrower cone and thus in the high energy part of the EXAFS spectrum multiple scattering becomes increasingly unimportant except when an approximately collinear geometry exists. Recently, the small atom approximation has also been modified to include multiple scattering contributions [41](Gurman and Campbell, unpublished).

These multiple scattering contributions and higher order scattering become even more important in the x-ray absorption near-edge region (XANES). Thus, the angular information of the ligands is contained in this region. Recent theoretical developments [42, 43] have made this region much more tractable and in the next decade we should expect an increasing number of applications for obtaining detailed stereochemical information.

Data analysis procedures have developed substantially over the last few years. In particular, use of least square refinement methods have been developed. Recent progress with theoretical development for the treatment of multiple scattering has resulted in ligand group refinement such as an imidazole. We can expect further development in this area which ought to lead us to restrained least square refinement procedures for EXAFS data analysis. This type of restrained refinement is commonly used for macromolecular crystallographic structure determination where a similar problem of underdeterminancy exists [44, 45].

4 Applications

During the last ten years XAS has been extensively used in probing the environment of metals in biological systems. The purpose of this section is not to review the impact of this technique in the field but to select a few examples to illustrate the type of study one may consider undertaking. Two groups of examples have been selected. The first group consists of proteins where the crystal structure of the native protein was known and XAS studies have either resulted in further refinement of the structure or a major correction to the crystal structure determination. In some of these cases XAS has been used to define the structural changes which may occur during a functional cycle for a protein such as reduction. The other group consists of proteins where no prior structural information was available and where XAS has made major contributions in defining the arrangement of the metal co-factor. A few examples, which deal with the study of intermediates, are also given. This is an area which is still not fully explored and much more work can be expected in the near future.

4.1 Iron-Sulphur Proteins

EXAFS has been extensively used to probe the prosthetic group of Fe-S proteins [46]. These proteins have attracted much attention in view of their importance as electron carriers. Rubredoxin, containing the minimum prosthetic group $Fe(SR)_4$, was the first protein to be studied by the EXAFS technique [47,48]. At the time its refined crystallographic structure was available [49], which had suggested that one Fe—S bond was extremely short, 1.95–2.05 Å compared to the other three Fe—S bonds at 2.3 Å. This strained structure was implied to be energetically poised to react easily. In view of its small size (MW = 6000 D), the EXAFS measurements were technically simple and the transmission method gave data of high quality. Shulman et al. [47] were able to show clearly that all the Fe—S bonds were equivalent and that the Fe—S distance was 2.26 ± 0.01 Å. Recent interpretations of crystallographic data have brought the two techniques in agreement [50]. This was therefore an important first Bio-EXAFS study which clearly demonstrated the complementarity of the two structural techniques. In addition, EXAFS data for the reduced protein showed an increase of 0.06 Å in the Fe—S distance. This information, most relevant to the function of this protein, is unlikely to have been obtained from crystallographic determination or any other presently available physical techniques with any certainty.

EXAFS has also been used for investigating a number of ferrodoxins (for a review, see Teo and Shulman [47]). As these studies were carried out a few years ago when rigorous analysis methods were not developed, only 2-shell analyses were performed and therefore information is available on only the Fe—S and Fe—Fe distances. A more comprehensive analysis should provide further details of the cluster. For ferrodoxins containing [2 Fe—2 S*] and (4 Fe—4 S*), Fe—S and Fe—Fe distances have been determined to be 2.25 ± 0.01 Å and 2.70 ± 0.03 Å. The increase in bond distances upon reduction for these proteins is small (<0.02 Å) and may just be considered outside the error limit. The small changes upon reduction are better defined compared to what may be possible from the crystallographic determination at 2 Å which had suggested a possible increase of 0.06 Å upon reduction [51].

A major controversy exists at present in the definition of the 3 Fe cluster by the EXAFS and crystallographic studies. Antonio et al. [52] have studied the iron K-edge EXAFS spectrum of the 3 Fe Ferrodoxin II of *Desulfovibriogigas* in the oxidised and reduced forms. For both forms they find that the Fe—S distance is 2.25 Å while the Fe—Fe distance is 2.7 Å. Both of these distances are consistent with crystallographic studies of model compounds and Fe—S proteins containing 2 Fe and 4 Fe centres. Stout and co-workers [53] have carried out crystallographic studies of *Azotobacter Vinelandii ferrodoxin I*, which have now been refined to 2 Å [54]. They find that the Fe—S distance is similar to that found in EXAFS studies, however the Fe—Fe distances are 4.18, 4.08 and 3.97 Å with a standard deviation of $\simeq$0.1 Å. It is possible that both of these structures are correct and therefore possible for the 3 Fe centre. Alternatively, one of the structural studies is in major error. The data presented by Antonio et al. [52] in their paper clearly shows the Fe—Fe distance to be nearer 2.7 Å than 4 Å. We note that crystallographic refinement of *Azotobacter ferrodoxin* was much slower than one would expect. A total of 187 cycles of refinement were used with Hendrickson's refinement program PROLSQ [45]. The structure had incorporated 344 water oxygens which is a significant proportion of the total of approximately 900 protein atoms. For the total of 1199 atoms (including water oxygen), only a modest R value of 0.262 was achieved. It is obviously important that further higher resolution EXAFS as well as crystallographic data are obtained with the aim of establishing the apparent flexibility of the 3 Fe centres.*

4.2 Metallothionein (Cu, Zn and Cd K-edges)

Metallothioneins are an ubiquitous class of proteins with which the important roles of metal storage (Cu, Zn) and detoxification (Cd, Pb, Hg) are associated [55,56]. EXAFS studies have been carried out on a variety of metallothioneins including Zn_7, Zn_5Cu_2, Zn_6Cu_3, Cd_7 and Cd_5Zn_2 to show that metal atoms are exclusively coordinated to the sulphur atoms of the cysteinyl residues [57-62]. In the case of Zn_7 metallothionein, evidence for metal clustering has been provided. Most of these studies have recently been reviewed [60] and are therefore not reproduced here. Details of the EXAFS structural parameters are given in Table 1. When this information is put in the context of the three-dimensional structure it should help in understanding how this protein is able to accommodate a variety of metals in different quantities and perform its dual function of metal storage and detoxification.

At present two three-dimensional structures are available, one determined by the 2D-NMR for solution protein (Cd_7MT) [63-65] and the other using the conventional crystallographic method [66,67]. These are in conflict and major discrepancies exist between the two structures. In addition, in the recent crystallographic refinement [67] where metal-sulphur clusters were also refined, metal-sulphur bond distances are obtained which are in serious error with respect to the EXAFS determined distances despite the claimed accuracy. For example, the Cd—S distances show considerable variation from 2.5 Å: the average terminal ligand distance is 3.2 Å while the bridging

* *Note added in Proof*: Stout, Turley, Sieker and Jensen, PNAS have repeated the crystallographic structure determination and show that the original crystallographic structure was in error in that in correct space group ($P4_3$, 2, 2) was assigned instead of P4, 2, 2. The new structure appears essentially the same as that of Peptococcus aerogenes ferrodoxin.

Table 1. Structural parameters deduced from EXAFS for a number of metallothioneins [60, 103]. R denotes the distance of scattering atoms from the absorbing atom; σ^2 is a Debye-Waller factor and is equivalent to ΔR^2_{rms}.

Zn in Zn_7MT^a			*Cu in* $^6Cu:3\,ZnMT^a$		
Atom	σ^2 (Å)^2	R (Å)	Atom	σ^2 (Å)^2	R (Å)^2
4 S	0.005	2.33 ± 0.02	3 S	0.005	2.25 ± 0.02
1 S	0.001	4.1 ± 0.1			
1 Zn	0.002	5.0 ± 0.1			
1 Zn	0.003	5.2 ± 0.1			
Cu in 5 $Cu:Zn_2MT^b$					
Atom	σ^2 (Å)^2	R (Å)			
3 S	0.007	2.25			
Cd in $Cd_7MT/Cd_5Zn_2MT^a$			*Ag in* $Ag_{17}Cd_2MT^a$		
Atom	σ^2 (Å)^2	R (Å)	Atom	σ^2 (Å)^2	R (Å)^2
4 S	0.006	2.53 ± 0.02	2 S	0.006	2.40 ± 0.02
			1 Ag	0.010	2.88 ± 0.04
Pb in Pb_7MT^a			*Hg in* Hg_7MT^a		
Atom	σ^2 (Å)^2	R (Å)	Atom	σ^2 (Å)^2	R (Å)^2
2 S	0.008	2.65	3 S	0.008	2.42 ± 0.02

[a] Measurements at 77 K; [b] Measurements at 300 K

ligand distance is 2.7 Å. The crystallographic distances for the metal-S bond cannot be real unless one accepts that they are so different as a result of the crystalline packing forces and then are non-representative of the solution protein. As yet there is no known case where the solution and crystalline distances for a protein's metal-ligand distances differ by any significant amount. In view of the modest size of the protein (6000 MW, 61 amino acids), such a major discrepancy in the metal-S distances is difficult to understand. This, together with major disagreement with the NMR structure, raises doubts about the correctness of the present crystallographic structure. We note that Stout and co-workers [67] have refined the data to 2 Å resolution. After 168 cycles of Hendrickson's PROLSQ refinement [45], the R factor was 21.6%. Forty-one water oxygens were incorporated into the structure, and also refined. The B-values (thermal parameters) were also refined and an average B-value of 22 for the whole structure is obtained. Refinement of individual isotropic temperature factors gave a mean value of 35 for the metal atoms. The higher values of B for metals and the long metal-to-cysteine distances were interpreted in terms of loosely bound metals. EXAFS evidence for both solution and freeze-dried protein is that the average metal-to-sulphur distances in metallothioneins are not significantly different from those in the inorganic metal-thiolate clusters. The long metal-to-sulphur distance and higher B-values found in the crystallographic refinement are not consistent with the role which metallothionein plays in terms of metal storage.

4.3 'Blue' Copper Proteins

"Blue" or Type 1 copper centres comprise one of the three types of copper found in biological systems. The distinctive properties of this class of protein is an intense blue colour associated with a very high absorbance in the visible region (around 600 nm), an exceptionally small hyperfine splitting in the g// region of the ESR spectrum and a high redox potential [68–71].

A qualitative understanding of these features was provided in 1978 when the crystal structure of poplar leaf plastocyanin and *Pseudomonas aeruginosa* azurin appeared [72,73]. The Cu in these proteins appears to be coordinated by two histidines with a Cu—N bond length of 1.9 to 2.1 Å, and a cysteine with a bond length of 2.1 to 2.3 Å. Evidence for additional weak interactions has been provided [74–76], a methionein (Cu—S of 2.6 to 3.2 Å) and in the case of *Alcaligenes* azurin for a carbonyl from glycine (Cu—O of 3.1 Å). EXAFS studies of freeze-dried powders of plastocyanin [77] (and more recently of a single crystal [78]) and azurin from *Pseudomonas aeruginosa* have yielded data in reasonable agreement with the crystal structure determinations. However, in these EXAFS studies the long Cu—S_{met} could not be determined. We note that a contribution from S_{met} would interfere with the carbon atoms of the imidazole rings and the two can only be separated if imidazole is treated as a restrained unit. The use of curved wave formalisms may also be beneficial as the low K-region of the EXAFS can then be utilised.

The redox behaviour of plastocyanin and azurin from *Pseudomonas aeruginosa* has been studied in some detail as a function of pH. For azurin, it has been found that at low pH, it is at least 2–3 orders of magnitude more active with respect to electron transfer to cytochrome-C551 than at high pH [79]. In contrast, plastocyanin has been found to be redox inactive at low pH [80]. Major progress has been made in understanding this behaviour in terms of structural features through the use of NMR, EXAFS and crystallography. In the case of plastocyanin, the crystal structure [76] has been determined at two pH values and no significant changes have been observed within the accuracy of the method which is estimated to be close to 0.1 Å. For the reduced protein, crystal structure has been determined for six pH values [81]. The most significant difference among the structures of reduced plastocyanin at different pH or between them and the oxidised protein are concentrated at the Cu-site. In the reduced protein, at high pH (7.8), the only significant change is the lengthening of Cu—N bonds by about 0.1 Å. Thus, at this pH, plastocyanin satisfies the conditions for an efficient electron transfer as the structural rearrangement required in going from Cu(II) to Cu(I) is only slight. The structure at low pH shows substantial changes; the Cu—S_{met} bond is shortened by some 0.4 Å while the Cu—N bond is broken, thus providing a very stable trigonal geometry and hence explaining why at low pH plastocyanin is redox inactive. For azurin, on the basis of NMR [82,83] and low resolution crystal structure [74], it was suggested that upon an increase of pH above pH 7, the environment of Cu would change. However, EXAFS studies [84] have shown that no major change takes place around the Cu atom. The use of the curved wave approach along with the difference method have proved useful in identifying the Cu—S_{met} contribution. The only significant change at higher pH is an increased uncorrelated motion of S_{met} with respect to Cu, which may result in a reduced efficiency of the electron transfer pathway or even an alteration of the pathway, thus reducing the efficiency of electron transfer at high pH.

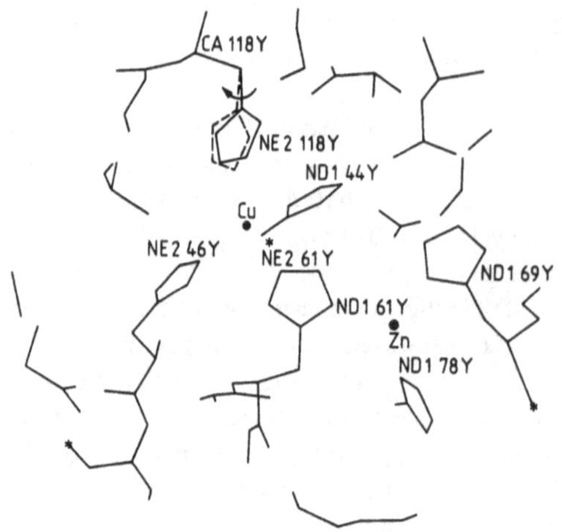

Fig. 6. Suggested changes in the coordination at the copper centre upon anion binding and during the catalytic cycle of Cu—Zn superoxide dismutase

4.4 Superoxide Dismutase (Cn-Zn)

XAS has proved particularly successful in defining the structural changes which occur upon reduction and anion binding to superoxide dismutase [85-90], thus providing a unique insight into the structure-function relationships in the Cu_2Zn_2SOD. The anion-binding studies have been particularly relevant due to the close similarity of the anions such as cyanide and azide to the superoxide substrate.

The early XAS study was carried out by Blumberg et al. [91] who showed that the Zn K-edge XANES did not change upon reduction, thus suggesting a very similar Zn environment for both forms of the protein. Blackburn et al. [86] extended the measurement to investigate both the Cu and Zn K-edges and found that even though the Zn K-edge data did not change, the Cu K-edge XANES changed dramatically. They also measured high quality EXAFS data for the Cu K-edge for both the oxidised and reduced protein solution and found that the overall amplitude of the Cu EXAFS had reduced by ~20% in the reduced protein. A rigorous analysis based on the single scattering curved wave method was carried out to define the distances of the imidazole groups accurately for both the native and the reduced protein. The change in distance (Cu—N = 1.94 Å (reduced) vs 2.00 Å (oxidised)) was also found to be compatible with the decreased coordination at Cu in the reduced protein.

Recent progress with the treatment of multiple scattering effects has proved valuable in distinguishing the ligands with similar or identical types of atom in the inner shell. For example, carbon or nitrogen atoms from the CN or N_3 ligand have been distinguished from the nitrogen atom of the imidazole [90]. This capability has been exploited for studying the structural changes which occur at the catalytic Cu site upon ligand binding such as CN and N_3. The structural similarity of these ligands to the superoxide substrate makes anion binding a valuable method of probing structure-function relationships in the Cu—Zn SOD.

Advances in the techniques of EXAFS analysis pioneered at the Daresbury Laboratory have allowed the unfiltered EXAFS data to be simulated accurately and the observed differences can thus be interpreted structurally with some confidence.

All double and triple scattering pathways which occur within the imidazole ring have been considered and this allows an accurate determination of all the imidazole atoms. In the case of the cyano-derivative, additional strong multiple scattering takes place due to the collinear geometry of Cu—C≡N. Multiple scattering treatment has allowed a refinement of the Cu—C≡N angle which comes out to 180° ± 5°. For the oxidised and cyano forms details of the outer shell distances are worth commenting on as these show splitting between C_2/C_5 and C_3/N_4 of about 0.2 Å. As has been noted by Blackburn and Hasnain [87], this splitting results from ring rotation. The EXAFS analysis package at Daresbury is being modified to incorporate the ring rotation. This development should allow a restrained refinement of a ligand group and therefore should minimise the number of parameters used. Also, this will provide more comprehensive geometrical information about the ligands.

From EXAFS alone, it is not possible to say which of the histidine imidazoles has moved away from the Cu atom when the enzyme is reduced. However, the fact that Zn K-edge data show no change when the enzyme is reduced, rules out the bridging ligand, His-61. This is more obvious when the three-dimensional structure is carefully looked at. A close inspection of the structure shows that histidine-118 is a ligand which can move most easily and thus would require minimum energy [Hasnain, unpublished]. A simple torsion of 30° around the C_A atom results in a lengthening of Cu—N_{His} to 2.7 Å, at which distance it is likely to make negligible contributions to the EXAFS data. As a result of this rotation, the imidazole ring from His-118 lies vertically above the Cu atom and provides sufficient space for a substrate or an anion such as CN to be accommodated at the Cu-site. Figure 6 shows the structural changes as deduced from XAS, molecular graphics and crystal structure. These structural changes are relevant to the mechanism of superoxide binding to Cu_2—Zn_2 superoxide dismutase.

4.5 Amine Oxidase (Cu)

Amine oxidase catalyses the oxidative deamination of amines to the corresponding aldehyde, hydrogen peroxide and ammonia. The copper containing amine oxidase from pig plasma (PPAO) is one of the better characterised in this class of enzyme. The homogeneously pure enzyme has a molecular weight of 190,000 composed of two subunits with equal molecular weight. The present evidence suggests that copper is essential for catalytic activity and therefore much effort has been made to determine the structure of copper sites [92–94].

Scott and Dooley [95] compared the EXAFS data for bovine plasma amine oxidase (BPAO) with that of $Cu(Imid)_4(NO_3)_2$ and concluded that the Cu^{2+} sites in native enzyme had 3–4 coordinated nitrogens (or oxygens) at 2.00 Å. Knowles et al. [96] have refined this model in several ways by making use of the curved wave theory for EXAFS, where multiple scattering effects have been taken into account. They have provided strong evidence for a thioether contribution at 2.38 Å, while the first shell distance has been found to be split. Their first shell analysis shows that primary scattering is due to Cu—O at 1.90 Å and Cu—N at 2.00 Å. Multiple scattering treatment shows that the Cu—N distance is compatible with two coordinated histidine imidazole ligands. Knowles et al. have proposed a model for the Cu-site, as shown in Fig. 7, which is consistent with the EXAFS data and earlier spectroscopic observations.

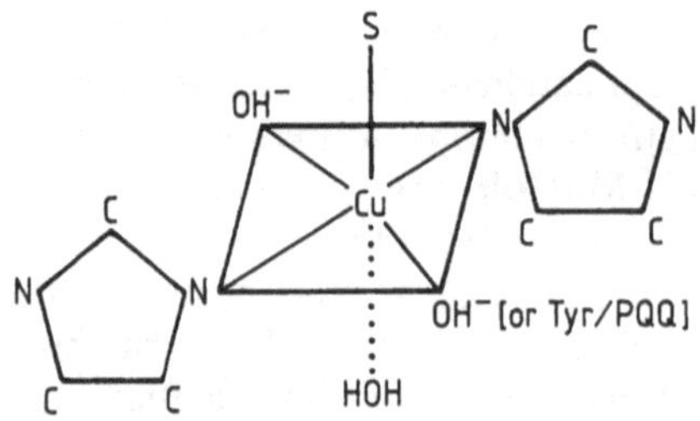

Fig. 7. A model for the Cu-site in amine oxidase proposed on the basis of XAS and other spectroscopic techniques

4.6 Angular Resolved XANES and EXAFS (Carboxymyoglobin Crystal)

We have seen above that XAS contains geometrical information about the back-scattering atoms through the multiple scattering effects. For a crystalline system, further sensitivity can be obtained, even when only single scattering is present. This is due to the large dichroism of XAS for crystals. In a crystalline system the EXAFS amplitude is given by

$$\chi_1(K) = \chi(K)\,[3\cos^2\theta]$$

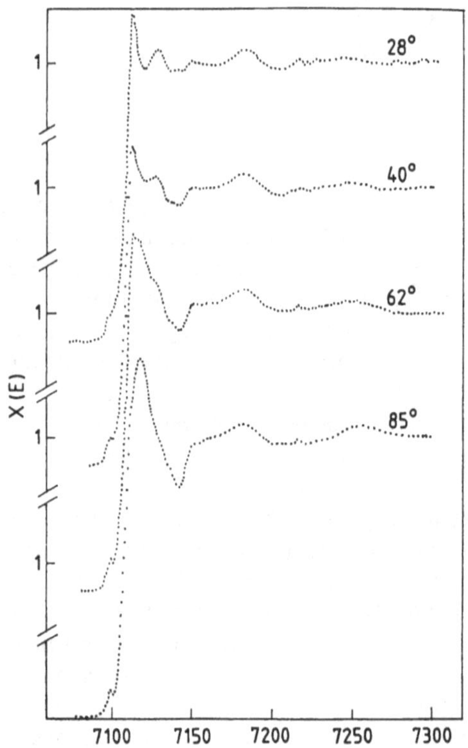

Fig. 8. Polarisation dependence of XAS data for carboxymyoglobin single crystal. For 28°, the contribution from Fe—CO is maximum while scattering from the porphyrin atoms is minimum. The spectrum for 85° is dominated by the scattering from porphyrin atoms

where θ is the angle between the electric vector $\vec{E}$ and the photoelectron scattering vector. Thus, the contribution for a particular ligand can be selectively enhanced by rotating the crystal to make an appropriate choice for θ. It is possible to enhance the contribution of a particular scattering centre by a factor of up to three. This aspect of XAS has been extensively exploited in studying adsorbates such as CO on metal surfaces [97]. However, the application of polarisation dependent scattering (or angular resolved XAS) to proteins has remained somewhat limited, mainly due to the lack of suitable crystals and to some extent due to the lack of a suitable analysis method which considered the multiple scattering effects. The recent development of multiple scattering EXCURVE formalism should encourage further applications of this powerful feature of XAS.

There are only two instances where polarisation dependence of XAS in a protein crystal has been studied [78,98]. A single crystal EXAFS study of poplar leaf plastocyanin has been carried out [78] in order to understand the absence of any significant contribution due to the Cu—S_{met} bond at ~2.9 Å. The other study was concerned with the carboxymyoglobin crystal, where it was primarily used to establish the suitability of this method for enhancing the contribution from Fe—C—O compared to that from Fe—N_p [98]. Figure 8 shows a polarisation dependent XANES and low K EXAFS region. The strong polarisation dependence is obvious. This study allowed an accurate determination of Fe—C distance and Fe—C—O angle. A more comprehensive experimental study with multiple scattering analysis could define the whole of the porphyrin cage very accurately. Such a study may be very valuable, particularly if a comparison with Mb and MbO_2 crystal is made.

4.7 Time-resolved Studies

The first successful time-resolved XAS study was accomplished by Mills et al. [99]. They studied the recombination of the CO ligand to myoglobin after laser flash photolysis. They were able to achieve a time resolution of 300 microseconds. They found that the recombination was complete in 1.2 ms. In view of the potential importance of such studies, it is worth noting some experimental details. A frequency doubled pulse from the Nd:YAG (neodymium:yttrium-aluminium-garnet) laser (15 ns FWHM, $\lambda/2 = 530$ nm) was synchronised with the x-ray pulse from the Cornell electron storage ring. A low concentration of 1 mM MbCO protein solution was used in order to achieve 95% photolysis. For a given x-ray energy, data were recorded during selected time intervals after the laser pulses and were stored in separate time bins. Thus, at a given x-ray energy, the spectra of both the photolysed and recombined states were collected, and thus any changes due to the source or instrument were eliminated. To reduce the effect of radiation damage, a flow cell was used with the flow speed chosen so that the entire sample volume exposed to the x-ray beam was renewed after several laser pulses. Mills et al. found that the largest source of noise was due to the Compton scattered radiation from the protein solvent, as is normally the case for static data for protein. In this case, the use of an energy dispersive multi-element detector array should provide a major improvement. In the above experiment, the rate limiting step was due to the recovery time for the laser (50 ms) compared to the recombination time of nearly 1 ms for MbCO. Thus data collection was made only at about 2% efficiency.

Chance and co-workers [100] have designed a flow system where the protein is continuously pumped optically using a tungsten or xenon flash lamp (764 nm). Using continuous illumination for various times and temperatures, Chance et al. have observed three intermediate states upon MbCO photolysis. At 40 K, a state with a recombination rate constant of 2×10^{-3}/s has been identified from two slower states with rate constants of $\sim 10^{-5}$/s.

Bray [101], who has pioneered the "rapid freeze" method for studying fast kinetics, has recently used this method in conjunction with XAS [102]. The procedure involves quenching reacting solutions by squirting them into a cold imiscible solvent and allows the preparation of a series of samples, each frozen at a particular reaction time (from about 3 ms upwards). Bray and co-workers [102] used this method to study the structural changes at the Mo(V) centre during the turnover of the slow substrate lumazine with xanthine oxidase. These initial results have indicated that some re-interpretation of the structure proposed from e.p.r. studies is required.

5 Summary and the Future

XAS has provided a new tool for investigating the metal cofactors in a protein with greater accuracy than has ever been possible before. The specificity of the technique along with its applicability to both crystalline and solution protein makes it a unique structural tool for deciphering the details of the metal co-factors. Structural changes associated with the pH and/or temperature jumps, oxidation/reduction cycle and anion binding have already provided special insights into the functioning of a number of proteins and should yield unparalleled information for many more systems in the coming years. Recent progress with time-resolved and freeze quench measurements has very exciting implications for studies of intermediates and much promise for the coming years. The increased sophistication of experiments such as the study of intermediates via continuous flow, flash photolysis etc. will require more and more specialised provisions at the experimental stage.

The development of sophisticated x-ray monochromators and the advanced sources such as multipole wigglers on the next generation (multi-GeV) storage rings at Grenoble and Chicago should provide the desired improvement in flux and brightness and open up a whole new range of proteins to be looked at by XAS. The advances in analytical procedures should make the treatment of multiple scattering effects routine and thus XAS should routinely provide the geometrical as well as the radial information around a metal centre.

6 Acknowledgements

I would like to thank a number of my colleagues, particularly Drs. N. Binsted, G. Diakun, S. Gurman and R. Strange and Professor C. D. Garner. I also thank Professor P. J. Duke, Dr. D. J. Thompson and Professor Leslie Green for their encouragement. I am especially thankful to Miss Julie Johnson for carefully reading the manuscript.

7 References

1. Garner, C. D., Hasnain, S. S. (eds.): EXAFS for Inorganic Systems, Daresbury Laboratory Publication DL/SCI/R17 (1981)
2. Bianconi, A., Incoccia, L., Stipcich, S. (eds.): EXAFS and Near Edge Structure, Springer Verlag (1983)
3. Hodgson, K. O., Hedman, B., Penner-Hahn, J. E.: EXAFS and Near Edge Structure III, Springer Verlag (1984)
4. Lagarde, P., Raoux, D., Petiau, J.: EXAFS and Near Edge Structure IV, Jour. de Physique, Colloque C8, Vol. 1 and 2 (1986)
5. Lee, P. A., Citrin, P. H., Eisenberger, P., Kincaid, B. M.: Rev. Mod. Phys. *53*, 769 (1981)
6. Bordas, J.: in Uses of Synchrotron Radiation in Biology, (ed. H. B. Stuhrmann) Academic Press, N.Y. p. 107 (1982)
7. Powers, L.: Biochem. Biophys. Acta *683*, 1 (1982)
8. Doniach, S., Eisenberger, P., Hodgson, K. O.: in Synchrotron Radiation Research, (ed. H. Winick, S. Doniach) Plenum N.Y. (1980)
9. Hasnain, S. S.: Life Chem. Reports *4*, 273–331 (1987)
10. Kincaid, B. M., Eisenberger, P. A.: Phys. Rev. Lett. *34*, 1361 (1975)
11. Kossel, W.: Z. Phys. *1*, 119 (1920)
12. Lea, K. R., Munro, I. H.: in Application of Synchrotron Radiation to the Study of Large Molecules of Chemical and Biological Interest, (eds. R. B. Cundall, I. H. Munro) Daresbury Laboratory publication, DL/SCI/R13 (1979)
13. Matsushita, T., Ishikawa, T., Oyanagi, H.: Nucl. Instum. Meths. *A246*, 377 (1986)
14. Van der Hoek, M. J., Werner, W., van Zuylen, P., Dobson, B. R., Hasnain, S. S., Worgan, J. S., Luijckx, G.: ibid. *A246*, 380 (1986)
15. Dobson, B. R., Hasnain, S. S., Hart, M., Van der Hoek, M., van Zuylen, P.: J. de Phys. *C8*, 121 (1986)
16. Mills, D. M., Henderson, C., Batterman: Nucl. Instrum. Meth. *A246*, 356 (1986)
17. Beaumont, J. H., Hart, M.: J. Phys. E 7, 823 (1974)
18. Bonse, U., Materlik, G., Schroeder, W.: J. Appl. Crystallogr. *10*, 338 (1976)
19. Greaves, G. N., Diakun, G. P., Quinn, P. D., Hart, M., Siddons, D. P.: Nucl. Instrum. Meth. *208*, 335 (1983)
20. Mills, D. M.: ibid. *208*, 355 (1983)
21. Cowan, P. L., Hastings, J. B., Jach, T., Kirkland, J. P.: ibid. *208*, 349 (1983)
22. Matsushita, T., Hashizume, H.: in Handbook on Synchrotron Radiation, (ed. Koch, E. E.), North Holland p. 261–314 (1983)
23. Pettifer, R. F., Foulis, D. L., Hermes, C.: J. de Phys. *C8*, 545 (1986)
24. Cramer, S. P., Scott, R. A.: Rev. Sci. Instrum. *52*, 395 (1981)
25. Phillips, J. C.: J. Phys. E. Sci. Instrum. *14*, 1425 (1981)
26. Hasnain, S. S., Quinn, P. D., Diakun, G. P., Wardell, E. M., Garner, C. D.: ibid. *17*, 40 (1984)
27. Baines, J. T. M., Garner, C. D., Hasnain, S. S., Morrel, C.: Nucl. Instrum. Meth. *A246*, 565 (1986)
28. Baines, J. T. M., Garner, C. D., Hasnain, S. S., Morrel, C.: J. de Phys. *C8*, 163 (1986)
29. Sayers, D. E., Lytle, F. W., Stern, E. A.: in Advances in X-ray Analysis (ed. B. L. Henke, J. B. Newkirk, R. Mallett) *13*, 248 (Plenum, 1970)
30. Sayers, D. E., Stern, E. A., Lytle, F. W.: Phys. Rev. Lett. *27*, 1204 (1971)
31. Stern, E. A.: Phys. Rev. *B10*, 3027 (1974)
32. Lee, P. A., Pendry, J. B.: ibid. *B11*, 2795 (1975)
33. Ashley, C. A., Doniach, S.: ibid. *B11*, 1279 (1975)
34. Pendry, J. B.: in Low Energy Electron Diffraction, Academic Press, N.Y. (1974)
35. Perutz, M. F., Hasnain, S. S., Duke, P. J., Sessler, J. L., Hahn, J. E.: Nature *295*, 535 (1982)
36. Gurman, S. J., Binsted, N., Ross, I.: J. Phys. C *17*, 143 (1984)
37. Gurman, S. J., Binsted, N., Ross, I.: ibid. *19*, 1845 (1986)
38. Teo, B. K.: in ref. 2, page 11 (1983)
39. Cramer, S. P., Hodgson, K. O., Steifel, E. I., Newton, W. E.: J. Amer. Chem. Soc. *100*, 2748 (1978)

40. Blackburn, N. J., Hasnain, S. S., Diakun, G. P., Knowles, P. F., Binsted, N., Garner, C. D.: Biochem. J *213*, 765 (1983)
41. Pettifer, R. F., Foulis, D. L., Hermes, C.: J. de Physiq. *C8*, 545 (1986)
42. Natoli, C. R., Kutzlev, F. W., Misemer, D. K., Doniach, S.: Phys. Rev. *A22*, 1104 (1980)
43. Durham, P. J., Pendry, J. B., Hodges, C. H.: Solid State Commun. *38*, 159 (1981)
44. Hendrickson, W. A., Konnert, J. H.: Biophys. J. *32*, 645 (1980)
45. Hendrickson, W. A.: Methods in Enzymology **115**, 252 (1985)
46. Teo, B-K., Shulman, R. G.: Chapter 9 in Iron-Sulfur Proteins, (ed. Spiro, T. G), John Wiley, 1982
47. Shulamn, R. G., Eisenberger, P. M., Blumberg, W. E., Stombauger, N. A.: Proc. Natl. Acad. Sci. *72*, 4003 (1975)
48. Shulman, R. G., Eisenberger, P. M., Teo, B. K., Kincaid, B. M., Brown, G. S.: J. Mol. Biol. *124*, 305 (1978)
49. Watenpaugh, K. D., Sieker, L. C., Herriot, J. R., Jensen, L. M.: Acta. Crystallogr. Sect. *B29*, 943 (1973)
50. Watenpaugh, K. D., Sieker, L. C., Jensen, L. H.: J. Mol. Biol. *138*, 615 (1980)
51. Carter, C. W., Kraut, J., Freer, S. T., Alden, R. A.: J. Biol. Chem. *249*, 6339 (1974)
52. Antonio, M. R., Averill, B. A., Moura, I., Moura, J. G., Omre-Johnson, W. H., Teo, B. K., Xavier, A. V.: ibid. *257*, 6646 (1982)
53. Ghosh, D., Furey, W., O'Donnell, S., Stout, C. D.: ibid. *256*, 4185 (1981)
54. Ghosh, D., O'Donnel, S., Furey, W., Robbins, A. H., Stout, C. D.: J. Mol. Biol. *158*, 73 (1982)
55. Kojima, Y., Kagi, J. H. R.: Tends. Biochem. Sci. *3*, 90 (1978)
56. Vasak, M., Kagi, J. H. R.: Met. Ions. Biol. Syst. *15*, 213 (1983)
57. Garner, C. D., Hasnain, S. S., Bremner, I., Bordas, J.: J. Inorg. Biochem. *16*, 253 (1982)
58. Abrahams, I. L., Garner, C. D., Bremner, I., Diakun, G. P., Hasnain, S. S.: J. Am. Chem. Soc. *107*, 4596 (1985)
59. Abrahams, I. L., Bremner, I., Diakun, G. P., Garner, C. D., Hasnain, S. S., Ross, I., Vasak, M.: Biochem. J. *236*, 585 (1986)
60. Hasnain, S. S., Diakun, G. P., Abrahams, I., Ross, I., Garner, C. D., Bremner, I., Vasak, M.: Proc. Int. Meet. Metallothionein 2nd, ed. H. R. Kagi, Birkhäuser: Basel (1987)
61. Freedman, J. H., Powers, L., Peisach, J.: Biochem. *25*, 2342 (1986)
62. Bordas, J., Koch, M. H. J., Hartman, H. J., Wesser, U.: FEBS Lett. *140*, 19 (1982)
63. Frey, M. H., Wagner, G., Vasak, M., Sorensen, O. W., Neuhaus, D., Wörgötter, E., Kagi, J. H. R., Ernst, R. R., Wüthrich, K.: J. Am. Chem. Soc. *107*, 6847 (1985)
64. Braun, W., Wagner, G., Wörgötter, E., Vasak, M., Kagi, J. H. R., Wüthrich, K.: J. Mol. Biol. *187*, 125 (1986)
65. Vasak, M., Wörgötter, E., Wagner, G., Kagi, J. H. R., Wüthrich, K.: ibid. (in press)
66. Furey, W. F., Robbins, A. H., Clancy, L. L., Winge, D. R., Wang, B. C., Stout, C. D.: Science *231*, 704 (1986)
67. Collett, S. A., Stout, C. D.: Recuil (Netherlands) *106*, 182 (1987)
68. Malkin, R., Malmström, B. G.: Adv. Enzymol. *33*, 177 (1970)
69. Gray, H. B.: Adv. Inorg. Biochem. *2*, 1 (1980)
70. Gray, J. B., Solomon, E. I.: in Copper Proteins (ed. Spiro, T. G.) John Wiley, 1981, pp. 3–39
71. Adman, E. T.: in Topics in Molecular and Structural Biology, Metalloproteins, Vol. 1 (ed. Harrison, P.) McMillan, 1985, pp. 1–42
72. Colman, P. M., Freeman, H. C., Guss, J. M., Murata, M., Norris, V. A., Ramshaw, J. M., Venkatappa, M. P.: Nature *272*, 319 (1978)
73. Adman, E. T., Stenkemp, R. E., Sieker, L. C., Jensen, L.: J. Mol. Biol. *123*, 35 (1978)
74. Adman, E. T., Jensen, L. H.: Isr. J. Chem. *21*, 8 (1981)
75. Norris. G. E., Anderson, B. F., Baker, E. N.: J. Mol. Biol. *165*, 501 (1983)
76. Guss, J. M., Freeman, H. C.: ibid. *169*, 521 (1983)
77. Tullius, T. D., Frank, P., Hodgson, K. O.: Proc. Natl. Acad. Sci. *75*, 4069 (1978)
78. Scott, R. A., Hahn, J. E., Doniach, S., Freeman, H. C., Hodgson, K. O.: J. Am. Chem. Soc. *104*, 5364 (1982)
79. Silvestrin, M. C., Brunori, M., Wilson, T., Darley-Usmar, V.: J. Inorg. Biochem. *14*, 327 (1981)
80. Segal, M. G., Sykes, A. G.: J. Am. Chem. Soc. *100*, 4584 (1978)

81. Guss, J. M., Harowell, P. R., Murata, M., Norris, V. A., Freeman, H. C.: J. Mol. Biol. *192*, 361 (1986)
82. Adman, E. T., Canters, G. W., Hill, H. A. O., Kitchen, N. A.: FEBS Lett. *14*, 287 (1982)
83. Canters, G. W., Hill, H. A. O., Kitchen, N. A., Adman, E. T.: Eur. J. Biochem. *138*, 141 (1984)
84. Groeneveld, C. M., Feiters, M. C., Hasnain, S. S., Rijn, J. V., Reedijk, **0**., Canters, G. W.: Biochem. Biophys. Acta. *873*, 214 (1986)
85. Blackburn, N. J., Hasnain, S. S., Diakun, G. P., Knowles, P. F., Binsted, N., Garner, C. D.: Biochem. J. *213*, 765 (1983)
86. Blackburn, N. J., Hasnain, S. S., Binsted, N., Diakun, G. P., Garner, C. D., Knowles, P. F.: ibid. *219*, 985 (1984)
87. Blackburn, N. J., Hasnain, S. S.: in Biological and Inorganic Copper Chemistry, (eds. K. Karlin, J. Zubieta) vol. 2, 33 (Adenine Press, 1986)
88. Strange, R. W., Hasnain, S. S., Blackburn, N. J., Knowles, P. F.: J. de Phys. *C8*, 593 (1986)
89. Strange, R. W., Blackburn, N. J., Knowles, P. F., Hasnain, S. S.: J. Amer. Chem. Soc. *109*, 7157 (1987)
90. Blackburn, N. J., Strange, R. W., McFadden, L. M., Hasnain, S. S.: ibid. *109*, 7162 (1987)
91. Blumberg, W. E., Peisach, J., Eisenberger, P. L. M., Fee, J. A.: Biochemistry *17*, 1842 (1978)
92. Knowles, P. F., Yader, K. D. S.: in Copper Proteins and Copper Enzymes, (ed. R. Lontie) Vol. 2, 103 (1984)
93. Blaschko, H., Buffoni, F.: Proc. Roy. Soc. *B163*, 45 (1965)
94. Barker, R., Boden, N., Cayley, G., Charlton, S. C., Henson, R., Holmes, M. C., Kelly, I. D., Knowles, P. F.: Biochem. J. *177*, 289 (1979)
95. Scott, R. A., Dooley, D. M.: J. Am. Chem. Soc. *107*, 4348 (1985)
96. Knowles, P. F., Strange, R. W., Blackburn, N. J., Hasnain, S. S.: ibid. (submitted, also available as Daresbury Laboratory report DL/SCI/P561E, 1987)
97. Norman, D.: J. Phys. *C19*, 3273 (1987)
98. Bianconi, A., Congiu-Castellano, A., Durham, P. J., Hasnain, S. S., Phillips, S.: Nature *318*, 685 (1985)
99. Mills, D. M., Lewis, A., Harootunian, A., Huang, J., Smith, B.: Science *223*, 811 (1984)
100. Powers, L., Chance, B., Campbell, B., Friedman, J., Khalid, S., Kumar, C., Naqui, A., Reddy, K. S., Zhou, Y.: Biochemistry *26*, 4785 (1987)
101. Bray, R. C.: Biochem. J. *81*, 180 (1961)
102. George, G. N., Bray, R. C., Cramer, S. P.: Biochem. Soc. Trans. *14*, 651 (1986)
103. Abrahams, I.: Ph. D. Thesis, Manchester University (1986)

Structure, Dynamics and Growth Mechanisms of Metal-Metal and Metal-Semiconductor Interfaces by Means of SEXAFS

Dominique Chandesris, Pascale Roubin and Giorgio Rossi

Laboratoire pout l'Utilisation du Rayonnement Electromagnetique CNRS-CEA-MEN, Universite' Paris-Sud, 91405 Orsay, France

Table of Contents

Structural information on the atomic arrangements at the early stage of formation of metal-metal, metal-semiconductor interfaces and semiconductor-semiconductor heterojunctions is needed along with the determination of the structure of the electron states in order to put on a complete experimental ground the discussion of the formation of solid-solid junctions. Amongst the structural tools that have been applied to the interface formation problem Surface-EXAFS [1-4] is probably the best suited since the local configuration of the interface atoms can be directly measured independent from the presence of long range order or of well known bulk phases. Furthermore vibrational properties of monolayer interfaces can be measured by SEXAFS. We will discuss the applicability of SEXAFS to the interface formation problem. After a brief review of the merits of the technique, we will review selected interface studies: the chemisorption and epitaxy of a fcc Co monolayer on Cu(111) and Cu(110) with the analysis of the crystallographic and dynamical properties of the Co adsorbate, and the chemisorption stage and silicide nucleation stage for the Pt/Si(111) interface.

1 Introduction

A large fraction of the material science research, and an important chapter of solid state physics are concerned with interfaces between solids, or between a solid and a two dimensional layer. Solid state electronics is based on metal-semiconductor and insulator-semiconductor junctions, but the recent developments bring the interface problem to an even bigger importance since "band gap engineering" is based on the stacking of quasi two dimensional semiconductor layers (quantum wells, one dimensional channels for charge transport).

The problem of surface magnetism and of the magnetic properties of thin metal films supported, or epitaxially grown onto single crystal substrates is also developing with an open interest for the possible exploitation of magnetic supports for information storage.

The research in catalysis is still one of the driving forces for interface science. One can certainly add to the topics of interface physics the whole new field of interface problems that is about to spring out of the new high Tc superconducting ceramics, i.e. the fundamental problem of the matching of the superconducting carriers wavefunctions with the normal state metal or semiconductor electron states, the superconductor-superconductor interfaces and so on, as well as the wide open discovery field for devices and applications.

One of the limiting factors in the development of a global response of experimental physics to the interface problem is that the initial growth of formation of interfaces (which is the key problem) is elusive to a detailed crystallographic investigation, whilst being rather well "seen" from the electronic states point of view, by exploiting electron spectroscopies. The surface sensitivity which is needed to resolve the interface specific signals from the overwhelming contributions from the bulk substrate is obtained in electron spectroscopy by exploiting the short mean-free path for electrons in solids, in a rather wide energy range to which correspond favourable matrix elements for a variety of excitation processes (photoionization, Auger electron excitation, electron elastic and inelastic scattering . . .). For the crystallography part of the problem only LEED exploits this parameter, but being a diffraction technique, it is sensitive to the long range periodicity of the surface and interface "in plane" order, referred to the laboratory frame, and therefore lacks information on the local atomic arrangements around the adsorbate atoms.

A surface sensitive version of the EXAFS technique has been attempted ten years ago, and has proven to be successful in a large variety of surface chemisorption and interface formation problems. In the following we recall very briefly what makes SEXAFS different from EXAFS and what is the specific information that can be withdrawn from the SEXAFS data, and address the problems of metal-metal interface formation, and metal-semiconductor interface formation with detailed examples.

1.1 SEXAFS of Interface Systems

The problems that one can address with the SEXAFS tool while studying the growth and formation of metal/semiconductor and/or semiconductor/semiconductor interfaces [5,6] are:

1) The bond lengths between the metal (semiconductor) adsorbed atoms and the surface of a semiconductor (metal) single crystal substrate at the chemisorption stage. The typical accuracy for data with a good energy range (300–700 eV) and good signal-to-noise ratio is of 0.02 Å [3] on bond lengths; ±15% on coordination numbers in case of a single chemisorption environment. The accuracy can be just as good in the presence of two or more chemisorption phases if they differ in bond distance by more than 1 Å, or if the first neighbour specie is different and the relative phase shifts are much different. In case of multishell contributions to the same EXAFS oscillations the accuracy decreases somewhat, or even a lot in the worst cases.
2) The adsorption site, i.e. the chemisorption position of the adatoms on (within, below) the substrate surface, thanks to the polarisation dependence of SEXAFS. Often a unique assignment can be derived from the analysis of both polarisation dependent bond lengths and relative coordination numbers. [2,7] The relative, polarisation dependent, amplitudes of the EXAFS oscillations indicate without ambiguity the chemisorption position if such position is the same for all adsorbed atoms. More than one chemisorption site could be present at a time (surface defect sites or just several of the ideal surface sites). If the relative population of the chemisorption sites is of the same order of magnitude, then the analysis of the data becomes difficult, or just impossible.
3) The changes in local coordination of the majority specie (the substrate) at the surface (e.g. adsorbate induced reconstruction). It all depends on the detection mode and on the absolute surface sensitivity of the most surface sensitive detection mode. The way to proceed is to measure differential SEXAFS spectra on one adsorption edge of the substrate material, where the difference is made between the clean substrate and the exposed substrate, or between a bulk sensitive detection of the EXAFS, (TY, FY, high energy AEY) and a surface sensitive measure (low energy AEY). [8] Attempts have been made on this ground, but final reports have not yet appeared in the literature.
4) The local coordination of the minority specie (i.e. the adsorbate specie) in the reactive interdiffusion stage (precursor compound, unreacted clusters).
5) The local coordination of the majority specie (substrate material) in the new interface phases. The interdiffusion region is often heterogeneous from the point of view of the substrate material (different first nearest neighbour environments) than for the adsorbate, therefore considerations on the resolution of multiple phases also apply.
6) The local coordination of the minority specie in the nucleated interface compound at the stage of compound formation. If one well defined compound is nucleated, then second and higher neighbour distances can also be measured in favourable cases. The problem of multishells analyses becomes identical to the bulk EXAFS case.
7) The presence of more than one specie in metastable equilibrium at the interface. Still the accuracy of SEXAFS can be good if the two phases have first distances sufficiently different (> 1 Å). This is not the case if two silicide phases, for example, are formed. In this case the presence of more than one phase can only be inferred from the inconsistency of the coordination numbers with any one only of the silicide phases, but the relative amounts or any better details are lost. It can be quite easy

instead if one silicide phase coexist with metal clusters, or with a metal rich phase where the nearest neighbours are metal atoms instead of silicon atoms. In this case, even if the absolute distances are similar, the phase shift difference can be strong enough to separate the experimental frequencies and allow an accurate analysis.

8) Informations on the vibrational and electron mean free path properties. Such analysis is possible only if the interface phase is very well defined, and if temperature dependent measurements are done and compared.[9,10] Debye Waller effects can be tangled with ordering transformation of the interface phase as a function of temperature and so on. If a single phase interface with order at least to the second nearest neighbour is recognised, then a temperature dependent Debye Waller, and mean free path analysis can be attempted.

Why do we need to know bond lengths and coordination numbers with so high accuracy on interface systems?

The interface formation problem is a difficult one. The best proof is represented by the wide experimental and theoretical effort of the last twenty years which has set the frame of interface science, but has left still wide open the key questions on the Schottky barrier versus ohmic phenomenology of the metal-semiconductor junctions, the band gap discontinuity distribution on the heterojunctions, the role of surface and interface parameters in catalysis, the kinetic thresholds for intermixing, alloying and compound nucleation at a solid/solid interface and so on. The difference in bond length and coordination number for different silicide compounds of the same metal-Si binary system are small. Metal-rich silicides have first Me—Si distances which differ by a few 10^{-2} Å. Recognition of the exact silicide phase formed at the interface, or accurate definition of a interface-specific reacted phase are results of paramount importance for the overall understanding of the interface phenomenology, and for the meaningful comparison with spectroscopic data[6] and with theoretical calculations of the band structure[5] and of the thermodynamical properties. Again in the case of surface magnetism the knowledge of the epitaxial growth of the magnetic layer, the bond distances with the substrate and the relaxation of the interlayer distances with respect to the bulk metals are primordial to the analysis of the magnetism itself. It is definitively clear that the accuracy of SEXAFS is needed in the interface studies, and that progress in interface science is expected to come from high resolution and very accurate, microscopic scale studies.

2 Rudiments of the Technique

The SEXAFS technique is a particular surface sensitive detection mode applied to an EXAFS measurement, performed in a particular environment, the Ultra High Vacuum, which is needed in order to prepare and protect the atomic-scale cleanliness of surface and interface systems for the whole time length of the experiment. As such the technique borrows equipment and procedures from the surface science spectroscopies,[2] and exploitation procedures of the synchrotron radiation source from the conventional, non vacuum, EXAFS method.[1] The studies described below have been performed at the Laboratoire pour l' Utilisation du Rayonnement Electromagnetique (LURE) exploiting synchrotron radiation from the storage ring DCI with a Si(311) double crystal monochromator. A Be window isolated the experimental chamber (pressure 10^{-8} Pa) from the primary vacuum of the beam line.

The physics of EXAFS is described, in scattering theory, as the modulations of the X-ray absorption coefficient as a function of excitation energy which are due to the interference between the photoexcited electron and the backscattered amplitude due to the presence of scattering centres (the neighbouring atoms) in condensed matter. The scattering of the photoelectron wave is assumed to be due to the point scattering centres represented by the very localised charge stored in the core levels of the neighbouring atoms. In the SEXAFS (i.e. in the EXAFS) regime the photoelectron kinetic energy is so high (from 50–100 to up to 1000 eV) that the scattering power of the atomic charge is rather low so that only single scattering events are of importance, and therefore the EXAFS signal contains intrinsically informations on the two body correlation function (i.e. one backscattering event and interference on the excited atom site).[11] In general terms SEXAFS is the measure of the empty density of states for extended energy ranges starting at 50–100 eV above the Fermi level.

The surface sensitivity is ensured by detecting the decay products of the photoabsorption process instead of the direct optical response of the medium (transmission, reflection). In particular one can measure the photoelectrons, Auger electrons, secondary electrons, fluorescence photons, photodesorbed ions and neutrals which are ejected as a consequence of the relaxation of the system after the photoionization event. No matter which detection mode is chosen, the observable of the experiment is the interference processes of the primary photoelectron with the backscattered amplitude.

Detailed descriptions of all the SEXAFS detection techniques are aboundant in the literature.[2,3] The most popular insofar are the electron techniques: the measure of the secondary (or total) electron yield which is ejected from the sample as a consequence of the excitation and decay of the primary core hole ensures rather intense signals with constant escape depth set by the minimum kinetic energy detected (typically in the secondary tail) but low adsorbed/substrate contrast; the Auger electron yield allow the tunability of the surface sensitivity as a function of the final state energy of the particular Auger transition which is chosen. The relaxation of the primary core hole causes a cascade of Auger (and fluorescence) characteristic transitions. The probability of each secondary decay process is constant, since the Auger (and fluorescence) decays are transitions between the atomic discrete energy levels, it follows that a direct proportionality holds between the primary photoionization event, and the relative EXAFS modulation, and any of the Auger (and fluorescence) decays of the excited atom. This fact allows one to attempt to analyse the depth variations of the local environments of a diffused atomic specie for example. AEY also gives a better adsorbate to substrate contrast since the secondary signal from the substrate at the same energy of the adsorbate Auger peak may be relatively small, but a worse signal to noise ratio.

The interface studies insofar have been done in electron detection mode, AEY and TY, whilst fluorescence seems to be a promising complement, with applications in the special case of multilayer arrays.

3 What the SEXAFS Data Look Like

SEXAFS scans are X-ray absorption versus photon energy curves starting a few tens of eV lower energy than a characteristic absorption edge (50–100 eV below threshold) and ending at as high energy as meaningful (and possible) above the edge, possibly 500–1000 eV above the threshold, typically 300–700 eV above the threshold. Examples are given in Fig. 1, for the Co K-edge (7709 eV) EXAFS of bulk and interface Co/Si(111) systems. The spectra contain the photoabsorption from all the electron states of lower binding energy than the excitation energy, which contribute a sloping, continuous background, since, in general, all EXAFS oscillations from these states are reduced to less than the statistical noise by the very large photoelectron kinetic energies for such shallower states and the correspondent very low scattering power of the atoms, and by the absorption jump and EXAFS modulations due to the characteristic absorption edge which is met in the chosen energy interval. For an homogeneous, ordered, system the EXAFS oscillations could still be seen at 1000 eV above the edge if the only damping factor were the mean square relative displacements of the atoms (thermal vibrations, Debye-Waller-like term); practically a number of phenomena limit the data range which can be exploited for the analysis. The limitations are of two kinds: intrinsic to the optical behaviour of the sample, i.e. intrinsic to the SEXAFS technique, and extrinsic due to the presence of partially avoidable artefacts, and to the always improvable data quality (signal to noise ratio, signal to background ratio).

Intrinsic limitations are:

1) photoelectron elastic peaks from the adsorbate or the substrate, sweeping through the electron analyser window;
2) Bragg glitches represented by multiple spikes and dips in the spectrum;
3) Higher energy absorption edges from the adsorbed or substrate material.

For metal/semiconductor interfaces the limitations coming from 1) are not very severe since the cross sections for photoemission and the lifetime broadening of such deep state photopeaks along with the poor energy resolution of the photon source

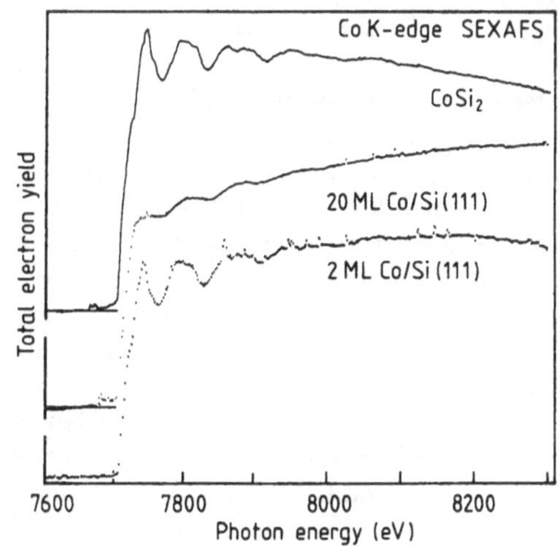

Fig. 1. Total electron yield (TEY) SEXAFS raw data on the K edge of Co overlayers on Si(111) 7×7 surfaces, and of $CoSi_2$. The bottom curve is for 2 monolayers of Co on Si(111) as deposited at room temperature. The reaction of Co with the Si substrate is seen from the similarity of the EXAFS oscillations with those of the $CoSi_2$ standard (top curve). The central spectrum was obtained from 30 ML Co/Si(111) as deposited at room temperature, and it is dominated by the EXAFS of unreacted Co. (Rossi et al. unpublished results)

(typically 3–10 eV for hard X-rays) contribute to smear the photoelectron elastic peak. Furthermore this problem can be drastically reduced or eliminated by adopting the TY or FY detection. The limitation coming from 2) are, on the other hand severe, in particular for high quality semiconductor single crystal substrates. The glitches, sharp peaks and dips in the SEXAFS spectrum, are due to abrupt changes in the effective excitation of the adsorbate and/or substrate atoms which arise from the occurrence of Bragg diffraction of the incoming photon beam on the crystallographic planes of the substrate crystal. Such diffraction events are unavoidable and depend on the wavelength of the X-rays and the relative orientation of the x-ray electric field vector and the crystallographic planes. One way to describe the phenomenon is to consider the interference between the incident beam (which in non Bragg conditions has a given optical path in the sample substrate) and the Bragg-diffracted beam. [12] The standing-wave field which results has peaks and nodes that travel within the crystal when the incident photon energy is varied (or the incidence angle of the light on the sample is varied), changing the effective excitation, i.e. the effective field intensity, in the region of the sample where the measured signal comes from, producing peaks and dips. The phenomenon is quite annoying for semiconductor, defect free, single crystals since the Bragg diffractions are coherent. Polycrystals or crystals with a high density of dislocations (like most metal single crystals) are less of a problem since the Bragg conditions are met over a broadened interval of X-ray energy and angular incidence; (of course this does not change the nature of the problem since instead of peaks one can have broad modulations of the photoexcitation, which could be misinterpreted as EXAFS oscillations). In our own experience the Bragg glitches are one major problem in the exploitation of the SEXAFS data from interfaces having the semiconductor as the substrate. The time consuming research of angular (both polar and azymuthal) tuning of the sample in order to move the glitches out of a reasonable energy window ($\geqq 300$ eV) above the edge is a leit-motive of the SEXAFS experimental runs.

The limitations due to the presence of a higher absorption edge from the adsorbate or from the substrate are rather a guideline to the choice of the proper way to do the experiment than actual limitations, unless the energy range of the synchrotron radiation outlet available is particularly short. For SEXAFS in the hard X-ray regime ($h\nu > 5$ KeV) the choice is often easy; in the soft X-ray range (100–4000 eV), on the contrary, if high angular momentum edges are sought, the total absorption spectra may become crowded with peaks and some edges cannot be exploited.

The extrinsic limitations are common to all spectroscopies:

1) signal to noise ratio. This limit the precision (reproducibility of the measures) and range of the SEXAFS data. Although at current levels of synchrotron radiation output intensities data precision can be as good as ± 0.01 Å [3] it takes long acquisition times which may introduces systematic errors like sample evolution and detection instabilities. Higher electron orbit stability in the storage rings and more intense photon fluxes are quickly becoming available, and are expected to continuously improve over the next decade. A better source will certainly prompt more applications and better quality data in SEXAFS on interfaces. Furthermore the strong bonds which are typically formed at semiconductor interfaces should not be significantly perturbed by more intense X-ray beams (one should think differently for the future of SAXAFS on organic or biological adsorbate systems . . .). At the current state of

the art SEXAFS of submonolayer interfaces still requires acquisition time of hours (1–6 hours) which, for unstable systems, or easily contaminated systems, means data averaging over more than one sample preparations.

2) The surface sensitivity is determined by the escape depth of the detected signal (elastic Auger detection) and/or by the penetration depth of the excitation (total reflection geometry) [13]. Practically the escape depth of the measure in Auger mode is determined by the electron analyser resolution. This is an important point since one is generally tempted to lower the analyser resolution in order to increase the count rate around a characteristic Auger peak. In doing so the actual measure is a sum of the elastic Auger peak (the only surface sensitive one), the partial electron yield tails on both sides of the peak (if the peak is 10 eV wide and one sets the analyser at 20–30 eV effective resolution the partial yield part of the count rate may be significant) and the partial yield background from the adsorbate and substrate underneath the elastic peak. Probably the best way of doing the measure is to normalise the AEY by a partial yield PY measure done with the same analyser at little different kinetic energy, like 20 or 30 eV shifted from the elastic Auger peak. This should assure that the background measured at those particular kinetic energies is properly normalised and that the truly surface sensitive signal is isolated. Such approach has not been done yet due to the relatively low Auger yield signals. The actual surface sensitivity of the state of the art AEY SEXAFS data is therefore worse than the ideal one, but significant improvements will come from higher photon fluxes and better analysers.

4 Rudiments of Analysis

The analysis of the SEXAFS data is basically identical to the analysis of conventional EXAFS data [1]. We will simply recall the basic ideas that sustain the conventionally used Fourier analysis of the SEXAFS data, making reference to Fig. 2, which will be further discussed below. The reason for being brief is that excellent reviews are available in widely diffused journals that were written by the promoters of the technique and warrant exhaustivity on the subjects [1–5]. The (S)EXAFS signal is defined as:

$$X(h\nu) = \frac{\mu(h\nu) - \mu_{atomic}(h\nu)}{\mu_{atomic}(h\nu)}$$

i.e. the oscillatory part of the X-ray absorption coefficient measured in the condensed matter system that deviates from the monotonic (in the hν range starting at ≈50 eV after the threshold energy) atomic absorption coefficient. The origin of the X-ray absorption spectrum is taken at the energy of the threshold for the core level excitation, at this energy the photoelectron has zero kinetic energy final state, above this energy the photoelectron wavevector is defined by:

$$k \doteq \sqrt{\frac{2m(h\nu - E_{binding})}{h^2}}$$

The EXAFS formula in the approximation of single scattering is the following:

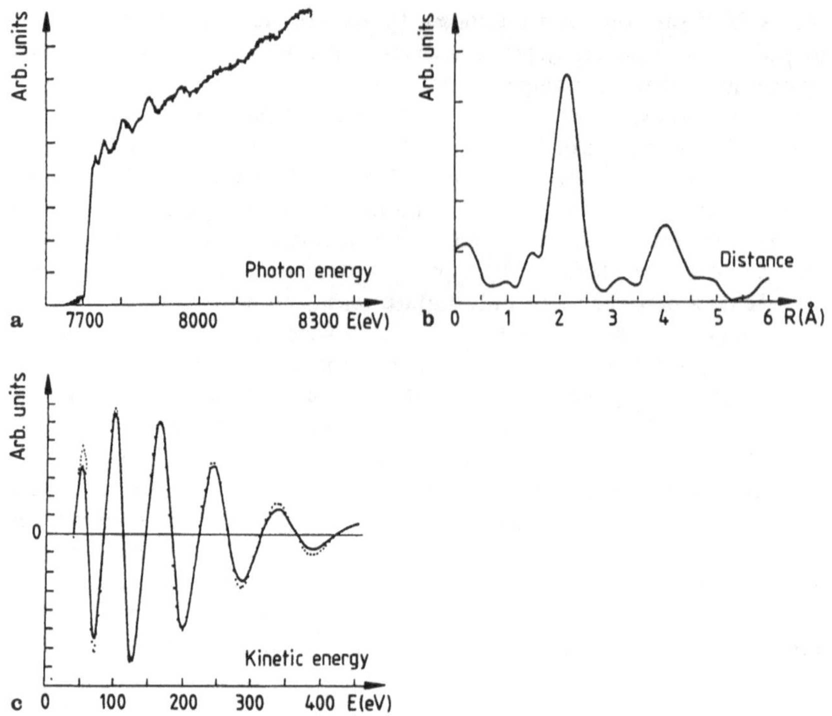

Fig. 2a–c. Steps of SEXAFS analysis applied to 0.3 ML Co/Cu(111) as deposited at room temperature. The curve **a** represents the raw absorption spectrum over at and above the Co K-edge. Curve **b** is the Fourier transformation of the background-subtracted EXAFS oscillations. The main peak represents the frequency corresponding to the first nearest neighbour contribution. Curve **c** is the inverse Fourier transformation of the first neighbour EXAFS signal extracted from the total absorption coefficient; it contains the information on the distance and (polarization dependent) coordination number. This information can be retrieved by fitting the curve with simulated EXAFS data for a proper geometry. The results are summarised in Fig. 3.[4,9]

$$X(h\nu) = -\sum_j \frac{N_j}{KR_j^2} e^{-\frac{2R_j}{\lambda(k)}} e^{-2\sigma_j^2 k^2} A_j(k) \sin\{2kR_j + \Phi_j(k)\}$$

which is the sum over all the neighbour shells j containing N_j atoms at the distance R_j from the absorber of the sinusoidal EXAFS contributions whose amplitude A_j is damped by the two exponential factors, and whose frequency is determined by the distance R_j and by the total phase shift undergone by the photoelectron wave during the scattering events [1]. Amplitude and phase shift are characteristic functions of the scattering pair (the excited atom, which, we know, and the backscattering atom at the distance R_j which is the unknown). Since in the adopted approximation the atoms of the solid are considered as point-scatterer only the localised charge, i.e. the core electrons, determines the $A(h\nu)$ and $\Phi(h\nu)$ functions. In other words the delocalised charge between the atoms does not contribute significantly to the scattering and therefore the $A(h\nu)$ and $\Phi(h\nu)$ functions are, so to say, chemistry independent.

This is a very fundamental hypothesis (very well verified indeed within the accuracy limits ±0.02 Å) of EXAFS and allows the analysis of an unknown system, say an interface between a transition metal and silicon, by using the amplitudes and phase shifts from a model compound of known crystallography, say a silicide.

The damping factors take into account: 1) the mean free path $\lambda(k)$ of the photoelectron; the exponential factor selects the contributions due to those photoelectron waves which make the round trip from the central atom to the scatterer and back without energy losses; 2) the mean square value of the relative displacements of the central atom and of the scatterer. This is called Debye-Waller like term since it is not referred to the laboratory frame, but it is a relative value, and it is temperature dependent, of course [14]. It is important to remember the peculiar way of probing the matter that EXAFS does: the source of the probe is the excited atom which sends off a photoelectron spherical wave, the detector of the distribution of the scattering centres in the environment is again the same central atom that receives the back-diffused photoelectron amplitude. This is a unique feature since all other crystallographic probes are totally (source and detector) or partially (source or detector) "external probes", i.e. the measured quantities are referred to the laboratory reference system.

One SEXAFS specific feature is the polarisation dependence of the amplitude. This derives from the high anisotropy of the surface and of ultrathin interfaces, that we may consider as quasi two dimensional systems. The relative orientation of the X-ray electric vector with respect to the surface (interface) normal does represent a preferential excitation for those atom pairs aligned along the electric vector; e.g. with the electric vector perpendicular to the surface (interface) plane the EXAFS amplitude will be maximum for the atom pairs aligned normal, or almost normal to the surface (interface). The electric vector can be also aligned, within the surface plane, along different crystallographic directions.

The EXAFS formula can be written as:

$$X(k) = \sum_i 3\cos^2 \alpha_i A_i^*(k) \sin[2kR_i + \Phi]$$

Table 1. First neighbour numbers (N) and apparent first neighbour numbers (N*) for linear polarized light giving, in normal and grazing incidence, the different contributions to the EXAFS signal in the case of a full monolayer coverage. S—S indicates Co—Co bonds between two surface atoms and S—B (S—B′) indicates Co—Cu bonds between a surface atom and a first (second) underlayer atom. φ is the angle between the projection of the electric field in the surface plane and the nearest-neighbour bond direction in this plane.

	Co/Cu(111)		Co/Cu(110)		
	S—S	S—B	S—S	S—B	S—B′
N	6	3	2	4	1
N* normal incidence ($\theta = 0°$)	9	1.5	$6\cos^2\varphi$ •	$3 + 3\sin^2\varphi$	0
N* grazing incidence ($\theta = 90°$)	0	6	0	3	3

where k is the photoelectron wave vector, and the summation runs over all the neighbours of the absorbing atom, so that i also indicates the bond between the absorbing atom and its ith bond direction. R_i is the mean length of this bond, and $A^{\infty}(k)$ is the total amplitude for the ith scatterer:

$$A_i^*(k) = [A_i(k)/kR_i^2]\, e^{-2k^2\sigma_i^2}\, e^{-2R_i/\lambda}$$

with the quantities as defined above. If one limits oneself to a first neighbour analysis has only two kinds of bonds i = S—S (surface—surface) and i = S—B (surface—bulk). One can then cary the contribution $3\cos^2\alpha_i$ by changing the angle of incidence of the synchrotron radiation beam onto the surface, as summarised in Table 1 for the Co monolayer/Cu systems.

The ability of probing surface anysotropic properties allows to define chemisorption sites based on purely geometrical considerations [2,7], as well as to investigate the anisotropy of the damping factors (Debye Waller-like) i.e. of the amplitudes of the relative atomic motions in the directions perpendicular to the surface, or in the surface plane [4,9,14].

5 Adsorption Geometry of 1/3 ML Co/Cu(111) and (1 × 1) 1 ML Co/Cu(111)

One third of a monolayer, and a full monolayer of Co adsorbed on the clean (1 × 1)Cu(111) surface were prepared, showing (1 × 1) LEED patterns [4,9,14]. According to previous results the growth mode is two dimensional, without Co—Cu mixing [15,16]. The SEXAFS data obtained for 1/3 ML Co/Cu(111), for energies larger than the Co K-edge ionisation are summarised in Fig. 2 [9]. Two kinds of neighbouring atoms are present for the Co adsorbate: Co atoms in the adsorbate plane, (S—S) and Cu atoms in the substrate (S—B). The Fourier transformations of the data show peaks corresponding up to the 4th nearest neighbours. At grazing incidence, (S—B bonds sampled) the first neighbour shell amplitude is similar for both 1/3 ML and full ML case, but at normal incidence (S—S bonds sampled) the ML amplitude is larger. This is consistent with the two dimensional adsorption mode. Indeed, in this case, the number of Co—Co bonds is the same at different steps of monolayer formation, whilst the number of Co neighbours in the adsorbate plane increases with the filling of all the chemisorption sites.

The results for the full monolayer are: Co—Cu (S—B) distance of 2.47 ± 0.03 Å and Co—Co (S—S) distance 2.51 ± 0.03 Å. This last distance is smaller than the Cu—Cu substrate distance, which is 2.55 Å, suggesting that some static disorder must be present in the adsorbate to allow pseudomorphism of the adsorbate adlayer [9], A good simulation of the data is obtained with an asymmetric Co—Co length distribution having 5 Co neighbours at 2.51 Å and one at 2.68 Å (corresponding to a mean distance of 2.54 Å). For 1/3 ML the Co—Cu distance is 2.51 ± 0.03 Å with an enhanced Debye Waller factor at RT (+10%). The in plane distances are 2.51 ± 0.03 Å, without static disorder (Fig. 3). The SAXAFS amplitude, in the adsorbate plane, for 1/3 ML is consistent with an average number of 3 nearest neighbours (instead

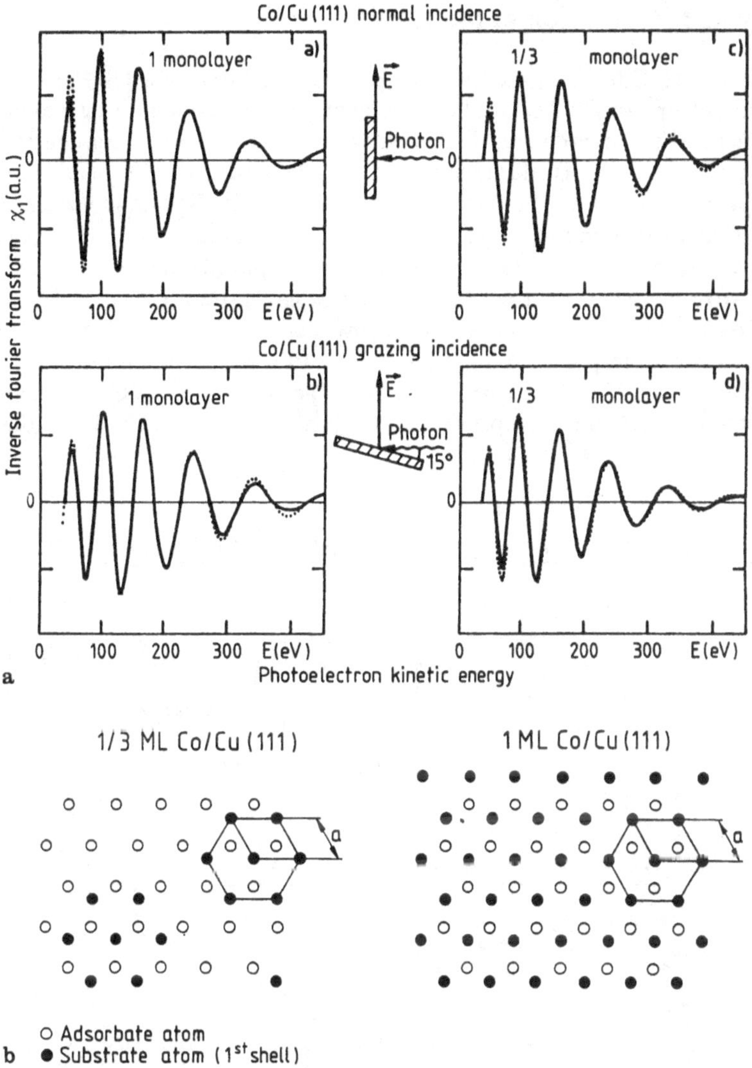

a

b

Fig. 3a. Filtered SEXAFS spectra (first shell) (......) and fits (———) for Co/Cu(111): a-one monolayer, normal incidence; fit: S—B bonds: 3 copper atoms (2.4 Å), S—S bonds: 5 cobalt atoms (2.51 Å) and 1 cobalt atom (2.68 Å) b-one third of a monolayer, normal incidence; fit: S—B bonds: 3 copper atoms (2.51 Å), S—S bonds: 3 cobalt atoms (2.51 Å), c-one monolayer, grazing incidence; fit: S—B bonds: 3 copper atoms (2.47 Å), d-one third of a monolayer, grazing incidence; fit: S—B bonds: 3 copper atoms (2.51 Å). **b.** Top view of absorbate geometrics compatible with the CdCu(111) SEXAFS results

of 6 for the full ML): this is consistent with the formation of small islands of 6–7 Co atoms. The coalescence of the islands at the monolayer cannot be done in perfect registry with the substrate due to the slightly shorter Co—Co distance, which explains the static disorder observed at 1 ML.

6 Geometry and Relaxation at the Co/Cu(110) Interface

The results for 1/2 ML Co/Cu(11), as deposited at 300 K are summarised in Fig. 4 [17]. The in plane anisotropy of the fcc(110) has been exploited by varying the azymuthal direction of the polarisation vector of the X-rays in the surface plane, as summarised in Table 2. The Co—Cu (S—B) distance is 2.46 ± 0.03 Å with the Cu neighbours in the top Cu plane, and also 2.46 ± 0.05 Å with the Cu atom in the second Cu plane. The Co—Co distance is found to be 2.51 ± 0.03 Å, i.e. the bulk Co distance, as in the case of adsorption onto the Cu(111) surface. The amplitude analysis indicates formation of two dimensional clusters of about three atoms. Data for 1.2 ML (Fig. 3) confirm the Co—Co and Co—Cu distances. A new Co—Co distance appears, due to the excess of Co with respect to a full monolayer, and is related to an incomplete second adsorbate layer (S—S′ = 2.45 ± 0.05 Å). The shorter Co—Co distance between the first and second (incomplete) adlayers can be understood as being due

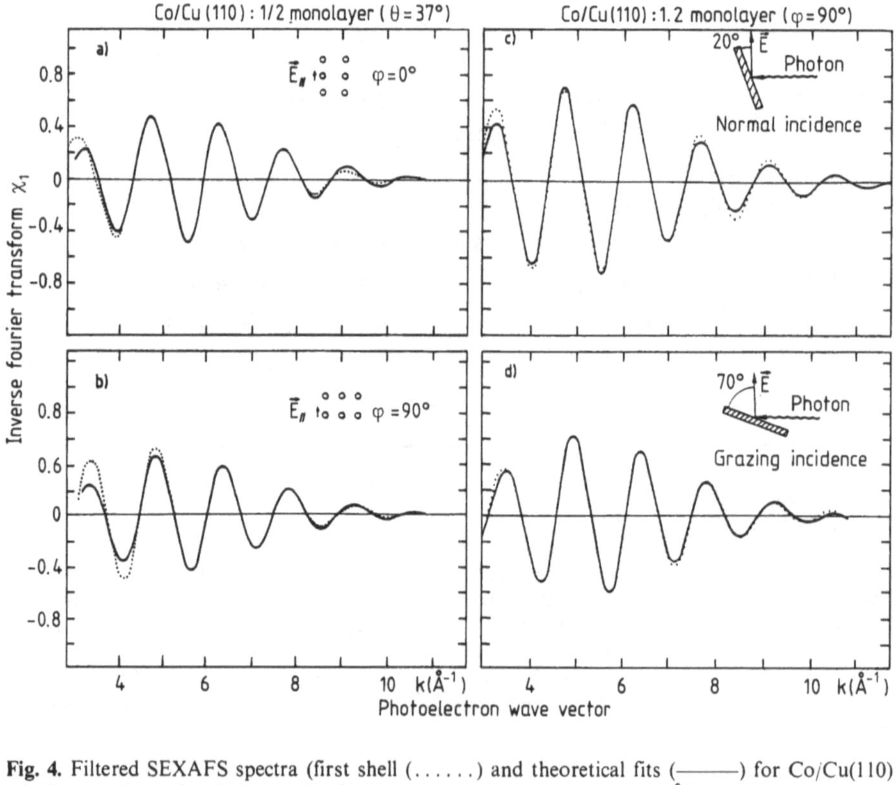

Fig. 4. Filtered SEXAFS spectra (first shell (......) and theoretical fits (———) for Co/Cu(110): a-half a monolayer, θ = 37°, Φ = 0°; fit: S—B bonds: 4 copper atoms (2.46 Å), S—B′ bonds: 1 copper atom (2.46 Å), S—S bonds: 1.3 cobalt atoms (2.51 Å), b-half a monolayer, θ = 37°, Φ = 90°; fit: S—B bonds: 4 copper atoms (2.46 Å), S—B′ bonds: 1 copper atom (2.51 Å), c-1.2 monolayer, Φ = 90°, normal incidence and d-1.2 monolayer, Φ = 90°, grazing incidence; fit: S—B bonds: 4 copper atoms (2.46 Å), S—B′ bonds: 1 copper atom (2.46 Å), S—B bonds: 2 cobalt atoms (2.51 Å), S—S′ bonds: 1.4 cobalt atoms (2.45 Å)

Table 2. Crystallographic analysis: experimental values of the incidence angle θ (between the direction of the light and the normal to the surface) and of the azimuth φ; related values N* of the different contributions, S—S, S—B, S—B′ for the systems Co/Cu(111) and Co/Cu(110) in the case of a full monolayer coverage.

	N*	S—S	S—B	S—B′
Co/Cu(111)	normal incidence θ = 0°	9	1.5	
	grazing incidence θ = 75°	0.6	5.7	
Co/Cu(110)	normal incidence θ = 20°	0	5.65	0.35
(φ = 90°)	grazing incidence θ = 70°	0	3.25	2.65
Co/Cu(110)	φ = 0°	3.8	3	1.1
(θ = 37°)	φ = 90°	0	4.9	1.1

to the reduced coordination for the topmost Co adatoms, which form islands of 2–3 atoms. A large amount of disorder is present in the 1.2 ML Co/Cu(110) system, which cannot be further analysed without a study of the temperature dependence of the SEXAFS amplitudes.

7 Vertical Relaxation at the Co/Cu Interfaces

On the basis of the SEXAFS derived bond distances one can attempt to derive the surface plane relaxation in the surface normal direction. By taking as a plausible reference a Co—Cu ideally non relaxed distance of 2.53 Å, intermediate between the bulk Co and bulk Cu, our experimental values lead to conclude that there is a contraction of this distance (Fig. 5). The contraction is found to be 4 ± 2% for the (111) face and 11 ± 5% for the (110) face. Moreover a small expansion of the second interlayer spacing (5 ± 5%) is measured in the (110) case. The oscillatory damping of the first interlayer distances has been observed before [18–20], and calculated, in particular in the case of Cu(110) [20–22].

The Co—Cu adlayer system is complex, but, due to the quasi epitaxy and negligible charge transfer, turns out to be a reasonable approximation to clean fcc surfaces.

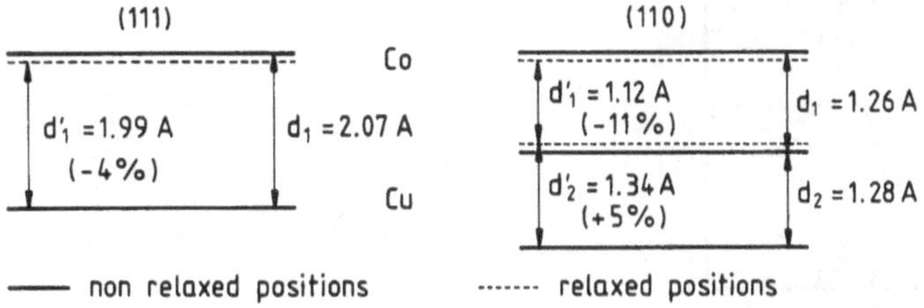

Fig. 5. Relaxation of the first interlayer distances for the (111) and (110) faces. An intermediate value R = 2.53 Å has been chosen as the eunrelaxed Co-Cu value. Relaxed interlayer distances, d_1, have been calculated for the two faces as: $d_1 = [d_1^2 + R_1^2 - R^2]_2^{\prime}$; d_1 is the unrelaxed interlayer distance and R_1 is the measured cobalt-copper distance, the copper atom being in the first underlayer. The (110) second interlayer distance is d_2': $d_2' = R_2' - d_1'$. If R_2' is the measured cobalt-copper distance, the copper atom being in the second underlayer. $[d_1' - d_1]/d_1$ is indicated in brackts (%)

Indeed phenomena typical of clean fcc surfaces as surface and subsurface relaxation are observed [23]. The (110) face exhibits larger effects than the (111) face because of its lower density.

8 Evidence of Anysotropic Dynamical Properties at the Co/Cu Interfaces

The quasi-ideality of the (1 × 1)Co/Cu(111) and (1 × 1)Co/Cu(110) monolayer interfaces allows a temperature dependent study of the polarisation dependent Debye Waller damping of the EXAFS oscillations: i.e: the analysis of the amplitude of the mean square relative displacements of the Co atoms parallel to the adsorbate layer, or perpendicular to it. The results are based on the analysis of data collected with the sample temperature T = 77 K and T = 300 K. The S—S and S—B (see above)

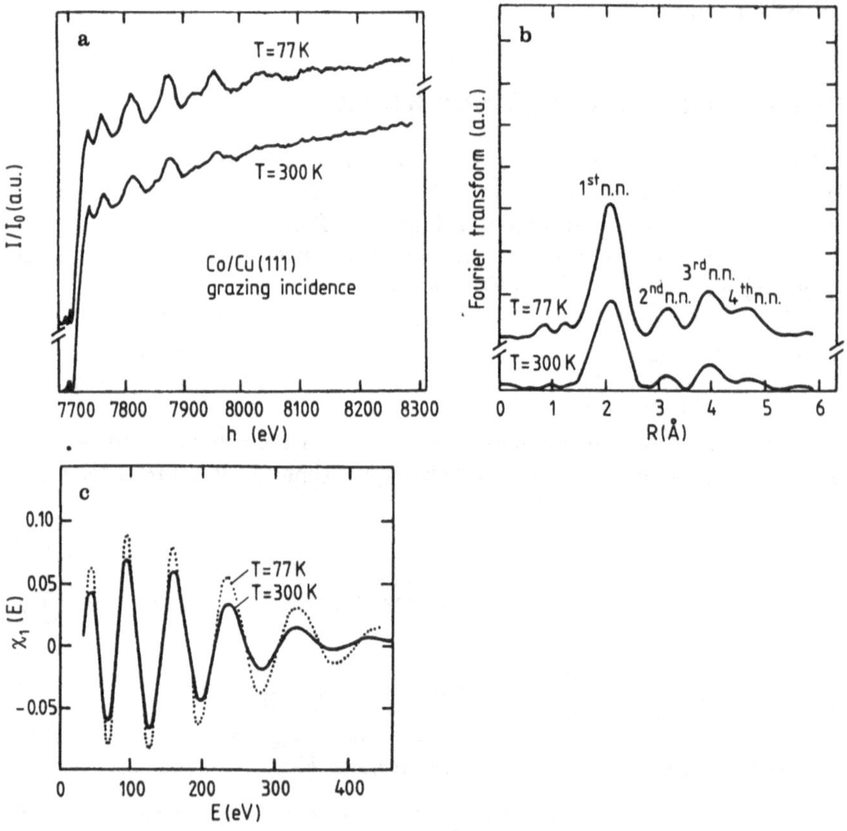

Fig. 6a–c. One monolayer of cobalt on copper (111), grazing incidence, T = 77 K and T = 300 K: relevant steps of the EXAFS analysis. **a** Experimental absorption spectra; **b** fourier transform of the EXAFS oscillations; **c** inverse Fourier transform of the first neighbour peak as a function of photoelectron kinetic energy E

Table 3. Thermal analysis: experimental values of the incidence angle θ (between the direction of propagation of the light and the normal to the surface) and of the azimuth φ; related values N* of the different contributions, S—S, S—B, S—B′, for the systems Co/Cu(111) and Co/Cu(110) in the case of a full monolayer coverage.

	N^*_{S-S}	N^*_{S-B}	$N^*_{S-B'}$
Co/Cu(111) normal incidence $\theta = 0 \pm 5°$	9	1.5	
Co/Cu(111) grazing incidence $\theta = 65 \pm 5°$	1.6	5.2	
Co/Cu(110) $\theta = 20 \pm 5°$, $\varphi = 90°$	0	5.6	0.35

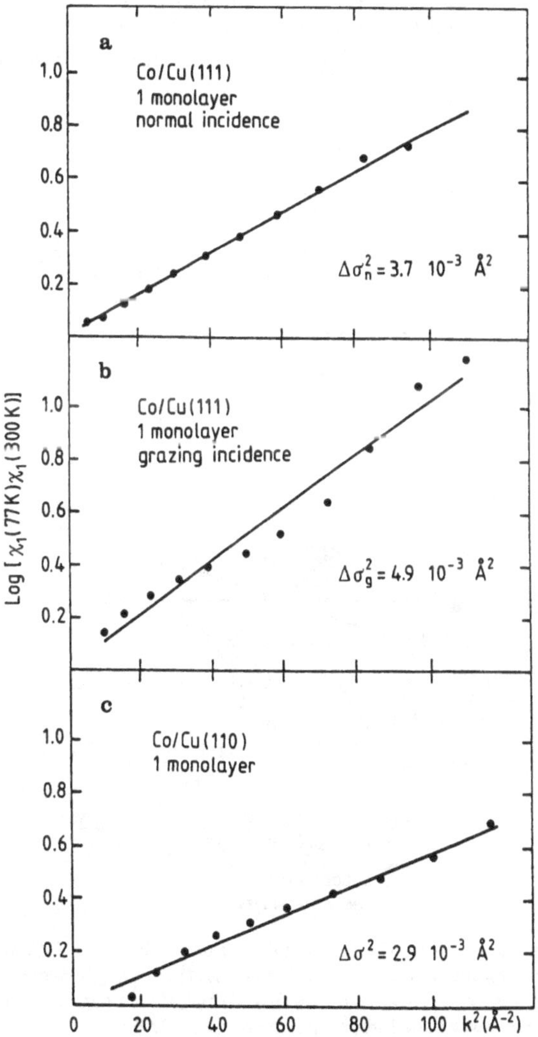

Fig. 7a–c. Plot of Log[X_1(77 K)/X_1(300 K)] as a function of k^2. Points are the experimental values of the logarithm calculated at the extrema of the inverse Fourier transform of the first neighbour peak. **a** One monolayer of cobalt on copper (111), normal incidence; **b** one monolayer of cobalt on copper (111), grazing incidence; **c** one monolayer of cobalt on copper (110). Continuous lines are the linear regressions corresponding to each spectrum. Related values of $\delta\sigma^2 = [\sigma^2(300\ K) - \sigma^2(77\ K)]$ are indicated

contributions are well separated in the different polarisation dependent spectra summarised in Fig. 6 [4, 9, 14, 17]. The analysis of each SEXAFS spectrum is done on the basis of the approximated formula:

$$X_1(T) = [N^*/kR^2]\ A\, e^{-2k^2\sigma^2(T)}\, e^{-2R/\lambda} \sin[2kR + \Phi]\,.$$

Where N^∞ is the effective number of bonds along the normal and in plane directions, as tabulated in Table 3. From this expression, the variation of the Debye-Waller factor between 300 K and 77 K is derived by the ratio method:

$$\mathrm{Log}[X_1(77\,\mathrm{K})/X_1(300\,\mathrm{K})] = 2k^2\Delta\sigma^2\,,$$

where $\Delta\sigma^2 = [\sigma^2(300\,\mathrm{K}) - \sigma^2(77\,\mathrm{K})]$. The formula is justified because factors other than σ^2 in the EXAFS formula are identical for the two temperatures (thermal expansion can be neglected). This empirical procedure eliminates the problem of disentangling the static disorder from the dynamical disorder, since, by cooling down the interface to

Table 4. Experimental values of $\Delta\sigma^2 = [\sigma^2(300\ \mathrm{K}) - \sigma^2(77\ \mathrm{K})]$.

	Co bulk	Co/Cu(111) S—S	Co/Cu(111) S—B	Cu bulk	Co/Cu(110) S—B
$\Delta\sigma^2$ ($\times 10^3$ A^2)	3.6 ± 0.3	3.6 ± 0.4	5.7 ± 0.8	4.7 ± 0.3	2.9 ± 0.4

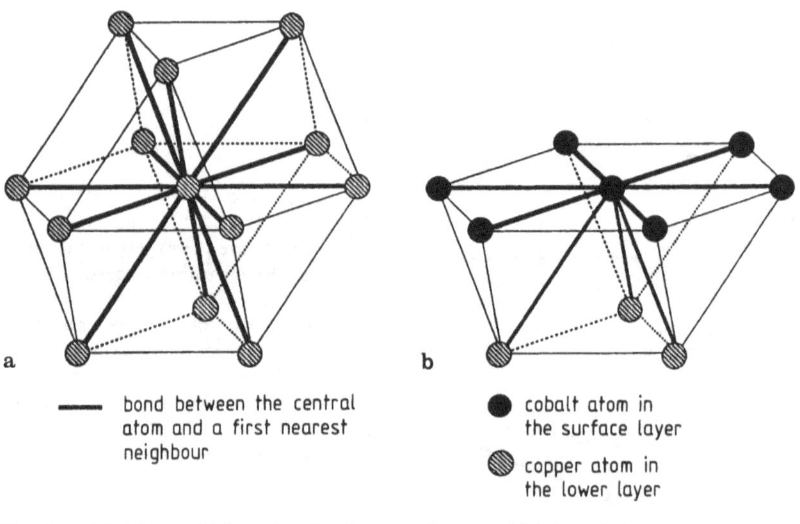

Fig. 8a and b. First neighbour bonds of a central atom which is: **a** bulk atom of a fcc crystal. It has 12 first-neighbour bonds. **b** A cobalt atom in a monolayer of cobalt on copper (111). In this case the three upper bond force constants are reduced to zero (surface effect) while the effect of the copper substrate is to reduce the values of the three lower bonf force constants by about 15%

77 K, the static disorder is certainly maintained. The results are presented as plots of formula in Fig. 7. The deviations from linearity of the plots is small enough to support such method of analysis. The slopes of the curves give the $\delta\sigma^2$ values tabulated in Table 4. It follows that in the (1×1)Co/Cu(111) case the anisotropy of surface vibrations clearly appears in the measured values of $\delta\sigma_n^2$ and $\delta\sigma_e^2$. There are two reasons for such anisotropy: the first is a surface effect due to the reduced coordination in the perpendicular direction. σ^2 is a mean-square relative displacement projected along the direction of the bond [17]. Enhanced perpendicular vibrational amplitude causes enhanced mean-square relative displacement along the S—B direction. The second effect is due to the chemical difference of the substrate (Fig. 8). S—B bonds are Co—Cu bonds and the bulk Co mean-square relative displacement, $\sigma_b^2(\mathrm{Co})$, is smaller than the bulk value for Cu, $\sigma_b^2(\mathrm{Cu})$. Thus for individual cobalt-copper bonds, the following ordering is expected:

$$\Delta\sigma_b^2(\mathrm{Co}) < \Delta\sigma(\mathrm{Co—Cu}) < \Delta\sigma_b^2(\mathrm{Cu})$$

therefore, even in absence of a surface effect, in the case of Co/Cu(111) we might expect $\Delta\sigma_{S-B} > \Delta\sigma_{S-S}$. A detailed analysis of the relative weight of the two effects is done in ref. [17]. The result is that the surface effect clearly dominates in the Co/Cu system, which can be well understood due to the similarity of the two kinds of atoms, the almost identical electronegativity, and therefore the negligible charge transfer between Co and Cu, which facts point to a similar atomic motion correlation factors.

The obtained $\Delta\sigma_{S-B^2} = 5.7 \times 10^{-3}$ Å^2 is even larger than the value of $\Delta\sigma_b^2(\mathrm{Cu})$ $\times (= 4.7 \times 10^{-3}$ Å$^2)$, and of the hypothetical Co—Cu crystal with intermediate elastic properties than bulk cobalt and copper $(4.1 \times 10^{-3}$ Å$^2)$. The derived effect of the effect of the lower coordination of the surface atoms on the mean-square relative displacement (perpendicular vs. parallel motions) is 1.4 times larger amplitude of the perpendicular vs. parallel motions, in agreement with lattice dynamics calculations. This SEXAFS study has produced a measure of the surface effect on the atomic vibrations. This has been possible due to the absence of surface or adsorbate reconstruction (i.e. no changes in bond orientations with respect to the bulk) and of intermixing.

8.1 Co/Cu(110): Stiffening of the Surface Force Constant

The Debye Waller analysis of the S--B bonds gives $\Delta\sigma_{S-B^2}(110) = 2.9 \times 10^{-3}$ Å^2. This value is lower than the pure Co value $(3.6 \times 10^{-3}$ Å$^2)$. Due to the low density of the (110) face, one might have expected a large mean-square relative displacement. The measured small value reveals a stiffening of the force constant of the Co—Cu bond. This is consistent with the large contraction of the Co—Cu interlayer distance ($\approx 11\%$; see above). The stiffening in strongly relaxed surfaces has been observed before [24-27] and overcompensate the effect of the reduced surface coordination in the perpendicular direction. Reversed surface anisotropy of the mean square relative atomic displacements has also been found on an other low-density surface: C2 × 2 Cl/Cu(110) {i.e. one half density of Cl vs. Cu(110) in plane density} where the Cl atoms moves with amplitudes parallel to the surface comparable with those of the Cu subtrate, but with a much reduced amplitude in the perpendicular direction [32].

9 Transition Metal-Silicon Interfaces: Chemisorption Sites and Silicide Growth

9.1 The Pt/Si(111) Interface Growth at Room Temperature

Total Electron Yield SEXAFS (and XARS) measurements were obtained on the L_3 edge of Pt submonolayers and monolayers deposited onto Si(111)7 × 7 at nominal room temperature conditions (Fig. 9) [28]. The analysis of the normal incidence spectra for 0.8 ± 0.2 ML Pt/Si(111) gives an unambiguous assignment of the sixfold interstitial chemisorption site between the top and second Si(111) layer. The Pt-(6 ± 1)Si bond length derived is 2.48 ± 0.03 Å, which is a typical Pt-silicide bond length (Pt_2Si = 2.46 Å; PtSi = average 2.5 Å). The deformation of the Si sixfold interlayer cage due to the Pt interstitial is discussed on the basis of the comparison of the 0.8 ML SEXAFS data and of simulated data which have been constructed using as inputs the experimental phase shift and backscattering amplitude and first neighbour distance (the one derived from the SEXAFS analysis) and higher neighbour distances corresponding to the various hypothesis of environment (Fig. 10). The sixfold interlayer site with vertical and lateral displacement of the six Si nearest neighbours, and smaller displacements also in the second coordination shell reproduce very well the experimental peaks, although a rather large error bar must be allowed for higher than first shell Fourier components (also due to the nois level seen at low K in the spectra). This is an important point: the changes in the substrate crystallography due to the presence of the adsorbate can be measured with SEXAFS either by directly measuring, in surface sensitive mode the substrate EXAFS, or by measuring with high resolution and a large K range the higher shell contributions in the adsorbate SEXAFS spectrum. The local coordination of Pt at the submonolayer stage is therefore basically identical to PtSi [Pt-(6) Si], apart the bond angles, but there is a very important difference: the SEXAFS spectrum is fully explained by all Si neighbours, i.e. non evidence of Pt second nearest neighbours is found, which makes the interface at this stage non silicide-like. This is an important point: the molecular orbital bonding structure between Pt and Si, with typical silicide bond length, is determined within the first cage, whilst the silicide-like electronic structure, which is also determined by the d–d interaction between Pt second neighbours, is lacking at the chemisorption stage, i.e. the ultrathin interface is not silicide-like [6]. For higher coverages the SEXAFS shows a dominant configuration for the diffused Pt in the substrate: Pt-(4 ± 1)Si at 2.46 Å. At 5 ML a second peak in the experimental Fourier transformation is only explained at least qualitatively, by a second neighbour shell of Pt atoms (Fig. 11). This corresponds to a local Pt_2Si configuration, and may represent the nucleation stage of the silicide within the Pt—Si solid solution produced by the intermixing. The Pt silicides do not grow epitaxially on Si(111), both the PtSi (orthorombic, Fig. 12) and Pt_2Si (tetragonal) have crystal structures of lower symmetry than Si and cannot nucleate starting from the diamond structure of the Si substrate. The nucleation within the disordered solid solution suggests a reaction phenomenology similar to a phase separation process. The evidence of such mechanism must be strengthened by other experiments, and by a careful kinetic study (in the described experiment the heating of the substrate during Pt evaporation was not controlled).

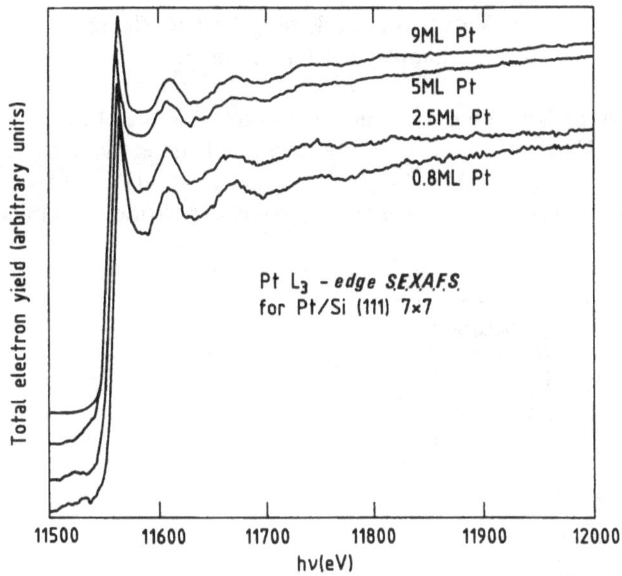

Fig. 9. Pt L_3-edge SEXAFS data obtained with TEY detection for Pt monolayers deposited at room temperature onto Si(111) 7×7 substrates (base pressure 1×10^{-8} Pa) [28]

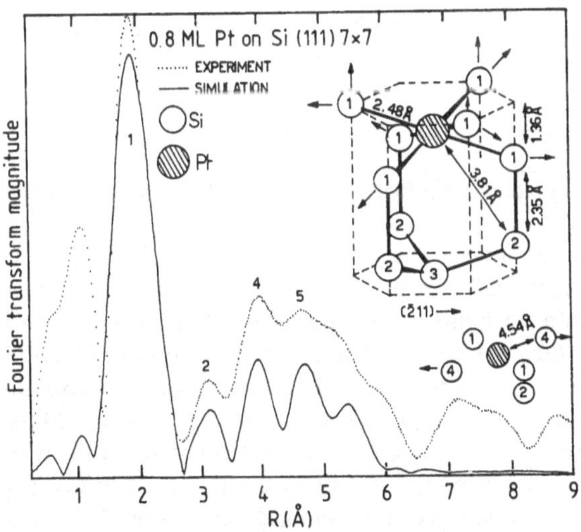

Fig. 10. Fourier transform of Pt L_3-edge SEXAFS data for 0.8 ML Pt/Si(111) 7×7 (dots) and of the simulated EXAFS for a Pt atom in the sixfold interstitial site within the top Si(111) double layer (solid line). The derived chemisorption model is represented in the insets. The first neighbour distance is Pt—Si = 2.48 ± 0.03 Å, and the Pt coordination number is 6 ± 1 Si neighbours. The distances beyond the first neighbour peak are fitted up to the fifth Pt—Si distance, and indicate the deformation of the Si cage, as indicated in the model, due to the presence of the host Pt atom [28]

The high resolution (better than 3 eV over the whole range) allows the discussion of the lineshape of the L_2 and L_3 edge resonances, which is discussed in the following paper.

One can safely remark that SEXAFS has made possible a direct picture of the local environment at the Pt—Si(111) interface. The sixfold interstitial site is the same as the Ni/Si(111) [29] {as well as Ag/Si(111) $\sqrt{3} \times \sqrt{3}$ [30], and Co/Si(111) [31]. Pt—Pt coordination starts to be seen only at 5 ML where the local coordination resembles

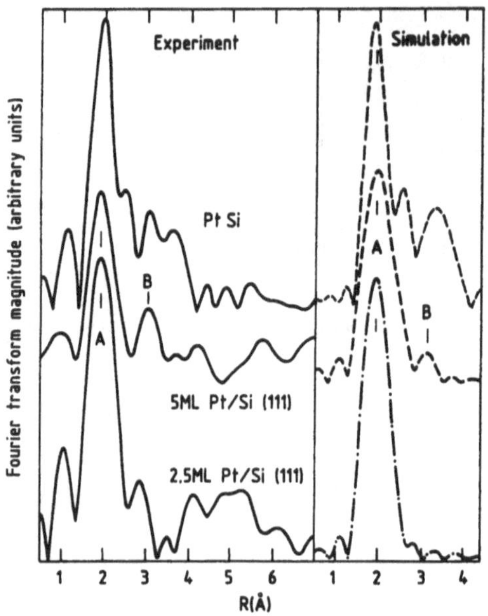

Fig. 11. Fourier transformations of PtL_3-edge SEXAFS data and simulations for 2.5 ML and 5 ML of Pt deposited onto Si(111), and PtSi standard silicide. The peak B is fitted by assming a local Pt_2Si coordination [28]

that of Pt_2Si suggesting a silicide cluster nucleation stage at the Pt/Si interface which follows the intermixing stage.

Similar results have been obtained for the early stage of growth of Ni/Si(111) [29] and Co/Si(111) [31], although an epitaxial growth of $NiSi_2$ (and $CoSi_2$) is observed instead of the mixing and nucleation mechanism. One can note that the chemisorption for the near noble metals, those which develop extended interfaces with the group IV semiconductors, develops as a strong interaction with the substrate surface, the deformation of the substrate top planes and the partial penetration into high coordination interstitial sites. The high coordination at bond lengths that are basically identical to those of the stable bulk phases is favoured independent from the fact that the geometrical cage be a precursor of the compound or not, and independent on the second shell coordination (which largely influences instead the electron states observed with spectroscopy). This uggests that the chemical bond structure between the near noble metal and silicon which is established at the chemisorption stage is basically the final one in terms of orbital hybridisation and of charge transfer. The ionicity versus bond length correlation based on Pauling electronegativity confirm this trend. The determination of the chemisorption interstitial position, its geometry, and the bond

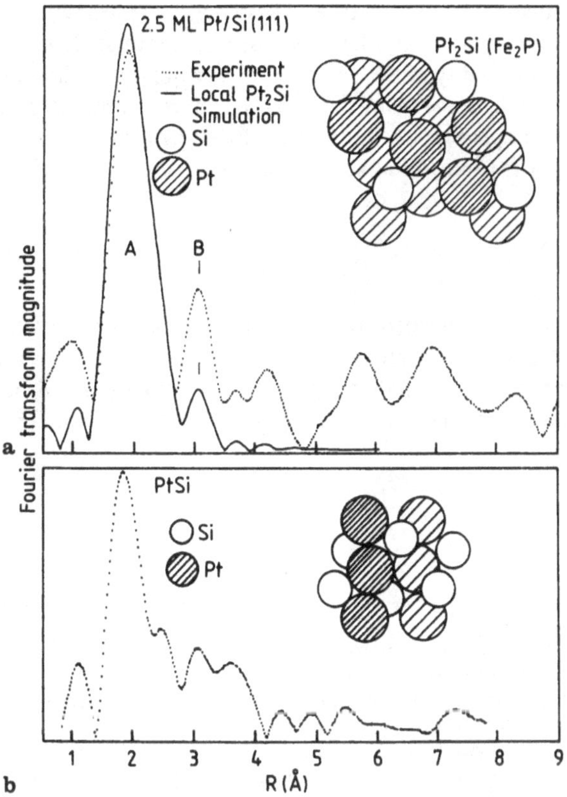

Fig. 12a and b. Fourier transform of EXAFS data, and structure model for **a** 2.5 ML Pt/Si(111) compared with a "local" (i.e. first and second shell) Pt_2Si configuration, **b** PtSi silicide

distances are very important inputs for theoretical calculations of the density of electron states and of the charge distribution at the ultrathin interfaces. Recent developments of PDOS calculation for silicide-like interfaces have modelled the bonding geometry by considering metal occupation of the adamantane interstitial cage of the diamond semiconductors. The sixfold subsurface interstitial site, and the deformation of the Si—Si bonds in the top Si(111) double layer are good explanations of the removal of the sp^3 hybridisation of the Si atoms which has been measured by Cooper minimum photoemission, and by Auger lineshape spectroscopy. The chemisorption stage can be therefore truly considered the precursor stage of silicide formation since only small bond angle variations intervene within the first shell and the silicide nucleation stage corresponds to the adding of the external shells of neighbours at the right distance and composition. Kinetic conditions and relative abundances of the atomic species may lead to different stoichiometries of the interface compounds.

SEXAFS is a powerful too for the crystallography of monolayer adsorbate systems; the applications discussed here on the problem of solid-solid interfaces show the key role that this technique plays in the development of the understanding of junctions, catalysis and surface magnetism.

SEXAFS insofar has been a time-expensive probe and systematic studies with the variation of all the interface growth parameters have not been attempted. It is not merely prosaic to foresee higher availability of SR beams of higher quality in the near future, and more clever detection arrangements. SEXAFS can certainly develop quickly towards a rôle of key technique for the study of the local coordination at all growth stages and conditions of a forming solid-solid interface.

10 Acknowledgements

The experiments described above were performed at LURE in collaboration with J. Lecante, whom we gratefully thank. The analysis and discussion of the surface vibrational anisotropies has been done in collaboration with G. Treglia and M. C. Desjounqueres. LURE is a French national facility, supported by the CNRS, CEA and MEN.

11 References

1. Lee, P. A., Citrin, P. H., Eisenberger, P., Kincaid, B. M.: Rev. Mod. Phys. *53*, 769 (1981), and references therein
2. Stöhr, J.: in X-ray Absorption: Principles, Applications Techniques of EXAFS, SEXAFS and XANES (ed. Prins, R., Koninsbergen, D.) Wiley, New York
3. Citrin, P. H.: Journal de Physique (Paris) C8, colloque 8, tome 47, (1986) p. 437, proceedings of the Intl. Conf. on EXAFS and Near edge Structure IV, Fontevraud, France 1986
4. Chandesris, D., Roubin, P., Rossi, G., Lecante, J.: Surf. Sci. *57*, 169 (1986); Chandesris, D.: Journal de Physique (Paris), Ecole d'Aussois de Rayonnement Synchrotron, Aussois, France 1986. (in french)
5. Calandra, C., Bisi, O., Ottaviani, G.: Surface Science Reports *4*, 271 (1985)
6. Rossi, G.: Surf. Sci. Rep. *7*, 1 (1987); Rossi, G.: in Semiconductor Interfaces: Formation and Properties (ed. Le Lay, G., Derrien, J., Boccara, N.): Springer Proc. in Physics 22, Springer Verlag, Berlin 1987, p. 69
7. Citrin, P. H.: Phys. Rev. *B31*, 700 (1985)
8. Comin, F., Incoccia, L., Lagarde, P., Rossi, G., Citrin, P. H.: Phys. Rev. Lett. *54*, 122 (1985)
9. Roubin, P., Chandesris, D., Rossi, G., Lecante, J., Desjounqueres, M. C., Treglia, G.: ibid. *56*, 1272 (1986)
10. Baberschke, K., Döbler, U., Wenzel, L., Arvanitis, D., Baratoff, A., Rieder, K. H.: Phys. Rev. *B33*, 5910 (1986)
11. Natoli, C. R., Benfatto, M.: Journal de Physique (Paris) *C8*, colloque 8, tome 47, (1986) p. 11; proceedings of the Intl. Conf. on EXAFS and Near edge Structure IV, Fontevraud, France, 1986
12. Ohta, T., Sekiyama, H., Kitajima, Y., Kuroda, H., Takahashi, T., Kikuta, S.: Jpn. J. Appl. Phys. *24*, L475 (1985)
13. Martens, G., Rabe, P.: Phys. Status Solidi (*a*) *58*, 415 (1980)
14. Chandesris, D.: J. Phys. (Paris) *C8*, 479 (1986)
15. Gonzalez, L., Miranda, R., Salmeron, M., Verges, J. A., Yndurain, F.: Phys. Rev. *B24*, 3245 (1981)
16. Miranda, R., Yndurain, F., Chandesris, D., Lecante, J., Petroff, Y.: Phys. Rev. *B25*, 527 (1982)
17. Roubin, P., Chandesris, D., Rossi, G., Lecante, J.: J. Phys. F, metal physics; (1987) to be published
18. Adams, D. L., Nielsen, H. B., Andersen, J. N., Stensgaard, I., Feidenshans, R., Sorensen, J. E.: Phys. Rev. Lett. *49*, 669 (1982)
19. Andersen, J. N., Nielsen, H. B., Petersen, L., Adams, D. L.: Solid State Phys. *17*, 173 (1984)

20. Frenken, J. W. M.: Thesis, Utrecht, The Netherlands; Gustfsson, T., Copel, M., Graham, W., Yasilove, S.: Bull. Am. Phys. Soc. *31*, 324 (1986)
21. Landman, U., Hill, R. N., Mostoller, M.: Phys. Rev. *B21*, 448 (1980)
22. Chen, S. P., Voter, A. F., Sroloovitz, D. J.: Phys. Rev. Lett. *57*, 1308 (1986)
23. Friedel, J.: "proc. Int. School of Physics Enrico Fermi, course LXI, Atomic Structure and Mechanical Properties of Metals", (ed. G. Gagliotti) North Holland 1976
24. Frenken, J. W., Van der Veen, J. F., Allan, G.: Phys. Rev. Lett. *51*, 1876 (1983)
25. Black, J. E., Franchini, A., Bortolani, V., Santoro, G., Wallis, R. F.: Phys. Rev. *B36*, 2996 (1987)
26. Treglia, G., Desjounqueres, M. C.: J. Phys. (Paris) *46*, 987 (1985)
27. Allan, G.: Surf. Sci. *89*, 142 (1979)
28. Rossi, G., Chandesris, D., Roubin, P., Lecante, J.: Phys. Rev. *B34*, 7455 (1986); J. Phys. (Paris) *C8*, 521 (1986)
29. Comin, F., Rowe, J., Citrin, P.: Phys. Rev. Lett. *51*, 2402 (1983)
30. Stöhr, J., Jaeger, R., Rossi, G., Kendelewicz, T., Lindau, I.: Surf. Sci. *134*, 813 (1983); Stöhr, J., Jaeger, R.: J. Vac. Sci. Technol. *21*, 619 (1982)
31. Comin, F., Citrin, P.: to be published; Rossi, G., Chandesris, D., Lecante, J.: unpublished results
32. Sette, F., Chen, C. T., Rowe, J. E., Citrin, P. H.: Phys. Rev. Lett. *59*, 311 (1987); Chandesris, D., Rossi, G.: Comment to the letter of Sette, F., et al.; Phys. Rev. Lett. *60*, 2097 (1988)

A Storage Phosphor Detector (Imaging Plate) and its Application to Diffraction Studies Using Synchrotron Radiation

Y. Amemiya[1], Y. Satow[1], T. Matsushita[1], J. Chikawa[1], K. Wakabayashi[2], J. Miyahara[3]

Table of Contents

[1] Photon Factory, National Laboratory for High Energy Physics, Oho, Tsukuba, Ibaraki 305, Japan
[2] Department of Biophysical Engineering, Faculty of Engineering Science, Osaka University, Toyonaka, Osaka 560, Japan
[3] Miyanodai Development Center, Fuji Photo Film Co. Ltd. Kaisei-machi, Ashigarakami-gun, Kanagawa 258, Japan

A new integrating area detector, originally developed for diagnostic radiography, has been applied to X-ray diffraction experiments using synchrotron radiation. The system is based on a photostimulable phosphor ($BaFBr:Eu^{2+}$) screen which can temporarily store an X-ray image. The stored image is read out by measuring the intensity of luminescence ($\lambda \sim 390$ nm) stimulated by a He-Ne laser beam scanning the screen surface. The area detector (250×200 mm^2) has more than 80% detective quantum efficiency for 8 ~ 17 keV X-rays, a dynamic range of $1:10^5$, a spatial resolution better than 0.2 mm (FWHM) in two orthogonal directions and no counting rate limitation. A high-quality diffraction pattern from a contracting muscle was obtained with a 10-s exposure. Laue diffraction patterns of cytochrome c′ and Weissenberg photographs of ω-amino acid: pyruvate aminotransferase were successfully recorded with an X-ray dose of approximately one thirtieth as much as that required by Kodak DEF-5 X-ray film. Small-angle scattering patterns from a synthetic polymer sheet during stretch were recorded in time-resolved mode.

1 Introduction

In order to make the best use of high-flux synchrotron X-rays in various kinds of diffraction and scattering experiments, it is necessary to develop an area detector which has a high quantum efficiency, a wide dynamic range and no counting rate limitation. Such an area detector is especially useful in experiments involving biological specimens where either the exposure time or the amount of the X-ray dose to the specimen is the most critical limiting factor.

Area detectors which function in a pulse-counting mode such as multiwire proportional counters, have been intensively developed [1-8] and used successfully in some applications. In other applications, however, the X-ray intensity available is completely beyond the counting-rate capability of these detectors, which can be only marginally improved. Therefore, integrating detectors such as X-ray film and X-ray TV detectors [9,10] are one of the logical choices for such application fields of synchrotron radiation. These detectors still have several problems. The main drawbacks of X-ray film are: 1) the dynamic range is limited (less than 2.5 orders of magnitude), 2) it has an "intrinsic chemical fog" level equivalent to about 0.15 optical density [11] which reduces the detective quantum efficiency, and 3) it is unsuitable for use in time-resolved experiments. For the X-ray detector the drawbacks are: 1) the size of the active area is relatively small, 2) the dynamic range is often insufficient (less than 3.0 orders of magnitude) and 3) it has spatial distortion and nonuniformity of response.

In this article, we describe a new type of integrating area detector system which has almost overcome the above drawbacks of conventional area detectors to meet the requirements in X-ray diffraction and scattering experiments using synchrotron radiation. This system was originally developed for diagnostic radiography [12,13]. In this system, an imaging plate, a photostimulable phosphor ($BaFBr:Eu^{2+}$) screen, larger than 250 mm × 200 mm is used to temporarily store a two-dimensional X-ray image as a distribution of color centers. The image is then read out by an image reader which measures the intensities of luminescence, which is released by the stimulation of a He—Ne laser beam scanning the surface of the phosphor screen (Fig. 1).

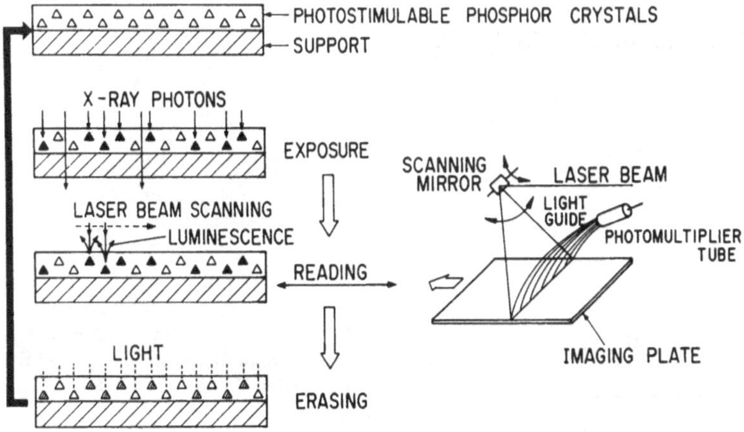

Fig. 1. Principles of image recording, reading and erasing with an imaging plate (Ref. [13])

In the following the principle and performance of the imaging plate is described; some of the results obtained so far using imaging plates are then presented. A new image reader system and a system which permits time-resolved measurements are proposed.

2 System Description

2.1 Imaging Plate (Photostimulable Phosphor Screen)

The photostimulable phosphor ($BaFBr:Eu^{2+}$) [14,15] shows a phenomenon called "photostimulated luminescence" (PSL). When exposed to X-rays, the phosphor temporarily stores a fraction of the absorbed X-ray energy in the form of quasistable color centers; when they are later stimulated by visible light, they emit photostimulated luminescence with an intensity proportional to the number of absorbed X-ray photons. The mechanism of the PSL can be explained by the energy level diagram shown in Fig. 2. By X-ray irradiation, a fraction of the Eu^{2+} ions are ionized to Eu^{3+} and the electrons which are liberated to the conduction band are trapped at quasi-stable levels of F-centers (a kind of color center). These levels originate from vacancies of Br^- ions in the phosphor crystal. By visible-light stimulation, the trapped electrons at F-centers are again liberated to the conduction band and return to Eu^{3+} ions, converting them to excited Eu^{2+} ions. Then, luminescence is emitted as a consequence of the electron transitions from the 5d to 4f shells in the Eu^{2+} ions. Figure 3 shows the spectra of the light which stimulates the phosphor and the spectra of the PSL [13]. The response time of the PSL is 0.8 μs. The crystal structure of the phosphor is shown in Fig. 4; some of the Ba^{2+} ions are replaced by doped Eu^{2+} ions.

The phosphor has the following characteristics which are suitable to obtain a high-quality image with a low X-ray dose: 1) the absorption coefficient for X-rays is high, 2) PSL radiation is efficiently released under such visible-light stimulation as a He—Ne laser beam, 3) the PSL-radiation spectrum falls within the range of

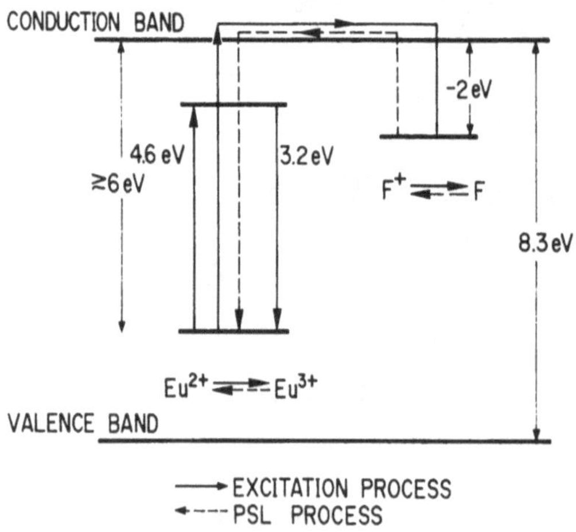

Fig. 2. The mechanism of photostimulated luminescence (PSL) in $BaFBr:Eu^{2+}$ (Ref. [14])

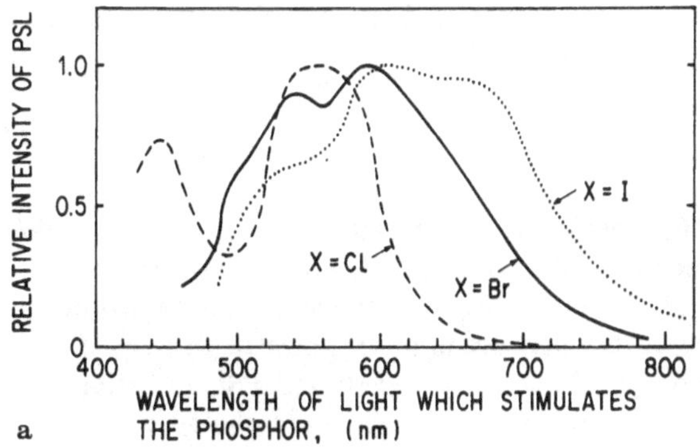

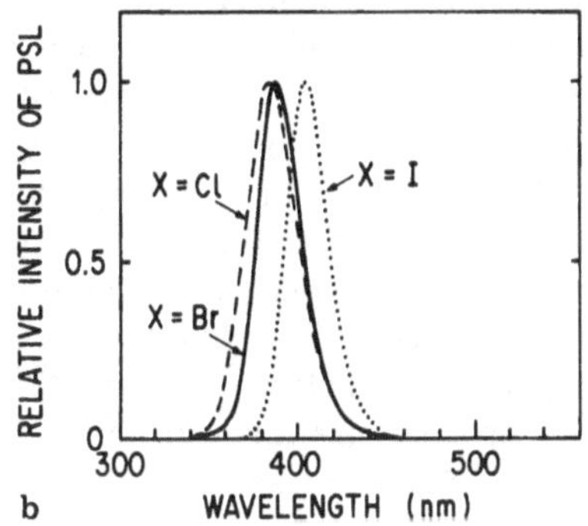

Fig. 3. **a** Stimulation spectra of $BaFX:Eu^{2+}$ (X = Cl, Br, I); **b** photostimulated luminescence spectra of $BaFX:Eu^{2+}$ (X = Cl, Br, I) (Ref. [13])

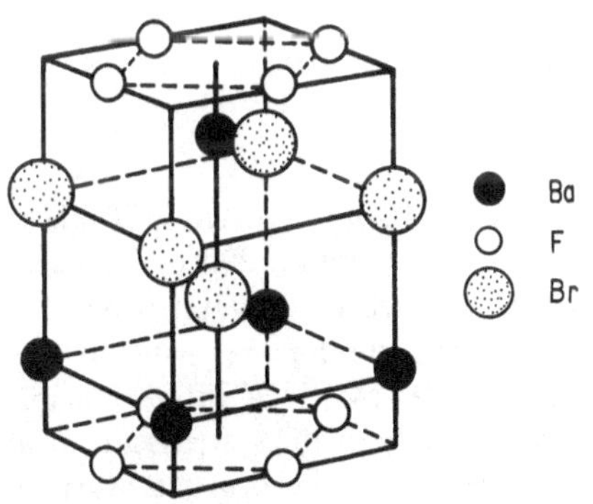

Fig. 4. A crystal structure of BaFBr (a = 0.451 nm, (c = 0.744 nm)

300–500 nm, where the quantum efficiency of photomultiplier tubes is high, 4) the wavelength of the PSL radiation (λ = 390 nm) is sufficiently different from that of the stimulating light for separate detection to be possible and 5) there is no substantial fading of the stored image for several hours.

The imaging plate (IP), approximately 0.5 mm thick, is a flexible plastic plate coated with polycrystals (crystal size: 4–5 μm in diameter) of the phosphor, typically 150 μm thick, combined with an organic binder. The surface of the phosphor layer is coated by a 10 μm thick polyethylene terephthalate sheet. A typical IP size is about 250 mm × 200 mm.

2.2 Image Reader

The X-ray image stored on the IP is read by a scanning He—Ne laser beam (20 mW) which releases PSL radiation. Scanning of the laser beam is performed with an oscillating mirror, while the IP is traversed so as to form an orthogonal scan (illustrated in Fig. 1). The PSL radiation released by laser scanning is collected through a light guide into a photomultiplier tube (PMT), which converts the PSL radiation into electrical signals. The output signals from the PMT, which are analog signals as a function of time, are logarithmically amplified and converted with an A/D converter into a time-series of digital signals (8 bits/pixel) that form histogram slices of the incident X-ray intensity on the IP. The PMT sensitivity and the amplifier gain can be preset for optimum reading. The scanning speed is about 14 μs per pixel. The minimum pixel size is designed to be 100 μm × 100 μm in the original system. The digital signals are sent to an image processor (a computer) for digital manipulation and are finally stored on magnetic tapes. The X-ray image is displayed on a photographic film by an image writer. In the image writer, another He—Ne laser beam modulated by the image signals is used to scan the photographic film to imprint the image. The residual image on the IP can be erased by irradiation with visible light of a normal light box to allow repeated use (Fig. 1). The IP can be used in almost the same ways as X-ray film, except that it can be treated in a light place before exposure to X-rays.

3 Detector Performance [16,17]

3.1 Active Area Size and Spatial Resolution

The active area size of the IP and the corresponding pixel size are listed in Table 1. All of the following description is concerned with an IP of 251 mm × 200 mm. Figures 5**a**), **b**) show intensity profiles on the IP when it is exposed to a 20 μm wide line-shaped monochromatic X-ray beam. X is along the laser scanning direction and Y is perpendicular to it. The full width at half maximum (FWHM) is less than 2 pixels (that is, 200 μm) in either direction. The observed intensity profiles are identical in several arbitrarily chosen regions within the active area. The spread in the intensity

Table 1. The area size of the imaging plate (IP)

active area size (mm × mm)	251 × 200	251 × 302	352 × 352	352 × 428
pixel size (mm × mm)	0.1 × 0.1	0.15 × 0.15	0.2 × 0.2	0.2 × 0.2
total pixel number	2510 × 2000	1670 × 2010	1760 × 1760	1760 × 2140

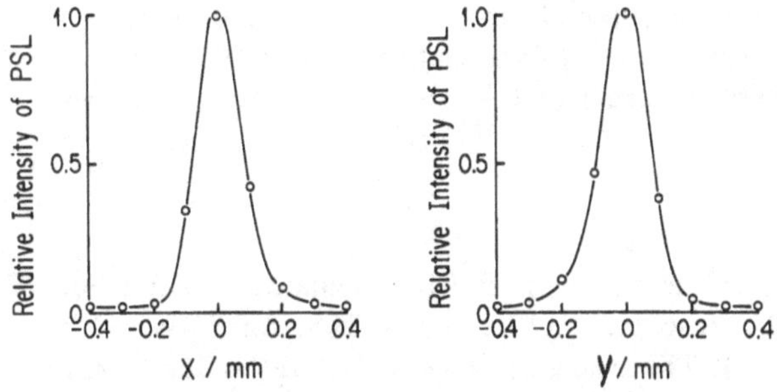

Fig. 5. Intensity profiles which show the spatial resolution of the IP along the laser scanning direction, X, and along the IP-scanning direction, Y

profiles is mainly due to the scattering of the laser beam in the 150 μm thick phosphor layer.

3.2 Linearity and Dynamic Range

The linearity and dynamic range of the PSL were measured with X-rays which ranged over six orders of magnitude. The IPs were read under three different sensitivities (E) of the image reader to cover the full intensity range. Figure 6 shows the relationship between the relative intensity of the PSL (y) and the incident X-ray photons/pixel (x) for both CuKβ and MoKα X-rays. According to a least-squares fit of the data to a straight line of the form, $y = a(x + b)$, the relative intensity of the PSL is linear to the incident X-ray intensity in the range from 8 to 4×10^4 X-ray photons/ pixel within a relative error of 0.05. The non-linearity between y and x is not negligible

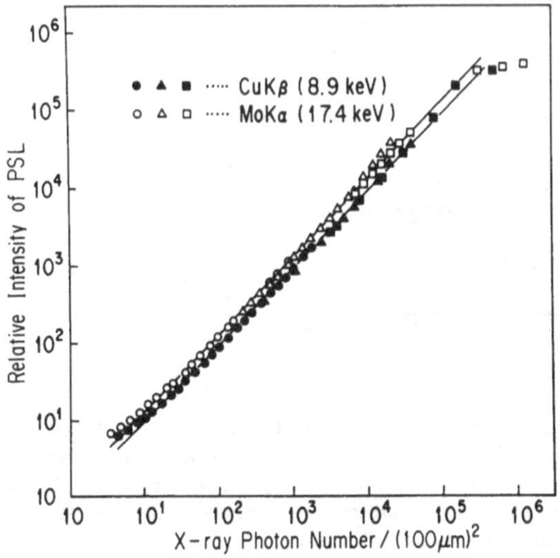

Fig. 6. The response of photostimulated luminescence to X-ray exposure level. ○, △, □ for MoKα, ●, ▲, ■ for CuKβ. IPs were read under three different sensitivities (E) (○, ●: E = 4000, △, ▲: E = 400, □, ■: E = 40) and 3.6 preset dynamic range

at higher exposure levels. A parabolic curve of the form $y = a\{(x + b) + c(x + b)^2\}$ fits the data over five orders of magnitude within a relative error of 0.05. The c value, i.e. the non-linear coefficient is of the order of $1 \sim 3 \times 10^{-6}$.

3.3 Efficiency

Figure 7 shows the absorption efficiency of the BaFBr calculated as a function of the X-ray photon energy. The absorption efficiency is 96% for 17 keV X-rays when the phosphor is 150 µm thick. The absorption edge at 37.4 keV is due to barium. The quantum efficiency of integrating detectors, however, can not be determined by the absorption efficiency alone, because the noise level of the system causes the quantum efficiency of these detectors to deteriorate. In Fig. 8, the relative uncertainty

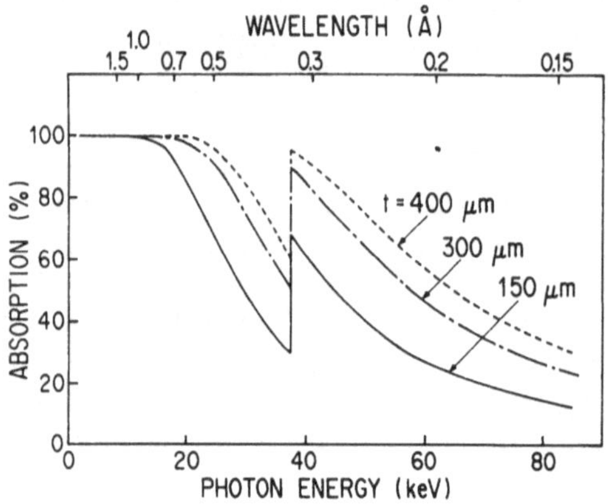

Fig. 7. Calculated X-ray absorption efficiency of BaFBr as a function of X-ray energy. t: thickness of BaFBr

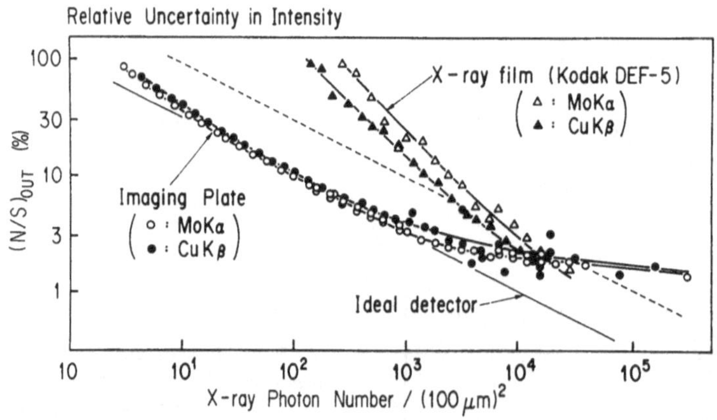

Fig. 8. The relative uncertainty in the measured intensity of various detectors as a function of the exposure level. A solid straight line indicates an ideal detector. A dashed line indicates a pulse-counting detector of 10% efficiency. ○ and ● indicate the IP system for MoKα and CuKβ, respectively. △ and ▲ indicate Kodak DEF-5 X-ray film for MoKα and CuKβ, respectively. The number of X-ray photons required to obtain a certain accuracy in intensity measurements can be compared

in the intensity is plotted as a function of the exposure level to allow comparison of the quantum efficiency of these dectors. The solid straight line indicates the relative uncertainty in the incident X-ray intensity; it follows Poisson statistics, where the quantum fluctuation is $\sqrt{n}$ for n photons. That is, if there exists an ideal detector, the results should fall on the solid straight line. From this graph, one can easily compare the incident X-ray photons which are necessary to obtain a certain desired accuracy in an intensity measurement; thus, the quantum efficiency can be compared. For example, the high-sensitivity X-ray film Kodak DEF-5 requires about 15 times and 25 times more exposure for CuKβ and MoKα, respectively, than would be needed by an ideal detector to obtain a 10% relative accuracy. The IP is nearly an ideal detector in the middle exposure range where the exposure level is between 10^1 and 10^3 photons/pixel. The deviation from an ideal detector at lower exposure levels is due to such system background noise, superimposed onto the output signal, as PMT dark current and scattered laser light which enters the PMT. The background noise of the IP amounts to less than 3 photons/pixel for both CuKβ and MoKα. This value should be compared with an intrinsic chemical fog level equivalent to about 1000 photons/pixel for X-ray film. The IP system also deviates from an ideal detector at higher exposure levels and the accuracy becomes saturated at around 2%. This deviation is partially due to errors of the apparatus used in the evaluation experiments and partially due to a non-linearity in the response of the PSL.

The efficiency of detectors can be discussed using a more general term called the detective quantum efficiency (DQE) [18]:

$$\mathrm{DQE} = \left(\frac{S_0}{N_0}\right)^2 \bigg/ \left(\frac{S_i}{N_i}\right)^2,$$

where S = signal and N = noise (standard deviation of signal) and subscripts o and i refer to output and input, respectively. The DQE of the IP system is obtained

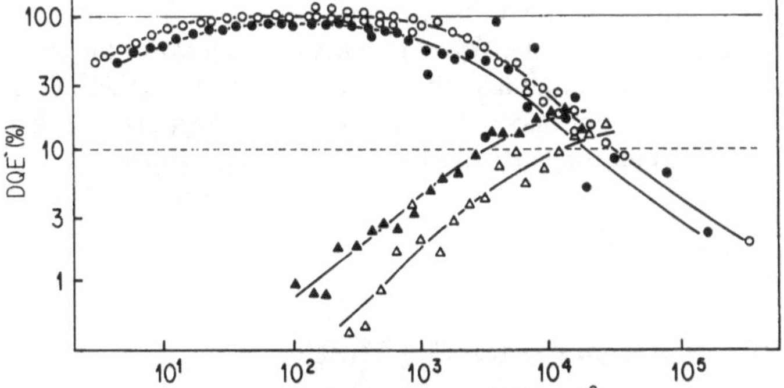

Fig. 9. Detective quantum efficiency (DQE) as a function of the X-ray exposure level. ○, ● indicate the IP system for MoKα and CuKβ, respectively. △, ▲ indicate Kodak DEF-5 X-ray film for MoKα and CuKβ, respectively

by comparing the relative uncertainty of the IP system with that of an ideal detector at the same exposure level. The IP system has more than 80% DQE in the exposure range between 10^1 and 10^3 photons/pixel. (Fig. 9)

3.4 Uniformity of Response

The uniformity of the response was measured by exposing a uniform X-ray beam (3.2×10^3 photons/pixel) of 10×10 mm^2 to 20 arbitrarily chosen areas. The integrated intensities of the PSL over the uniformly exposed areas were compared and the macroscopic non-uniformity defined by the relative uncertainty among the integrated intensities was 1.6%. This macorscopic non-uniformity may be due to non-uniform stimulation by the scanning laser beam deflected with an oscillating mirror in the image reader. The microscopic non-uniformity defined by the relative uncertainty among individual pixel intensities within the uniformly exposed area was 2.2%. If the incident photon fluctuation is taken into account, the intrinsic microscopic non-uniformity is estimated to be 1.3%.

3.5 Image Distortion

The IP was exposed to line-shaped monochromatic X-ray beams with exact separations. The ratios of the resulting separations to the true values were examined at several regions over the whole active area so as to evaluate the image distortion. It was 1.010 ± 0.001 and 0.996 ± 0.001 along the laser-scanning direction and the orthogonal direction, respectively. The image distortion is attributed to an error in the scanning mechanism of the image reader, and can be calibrated for accurate spatial measurements.

There is no appreciable deterioration in the performance of the IP due to radiation damage or any physical fatigue of the PSL phenomenon so far as the IP is used in the normal way. Therefore, the IP can be used many times unless it becomes mechanically damaged. However, an extraordinarily intense direct beam of white synchrotron radiation will produce color-centers which cannot be easily erased. The tolerance of the X-ray dose to the IP is now being quantitatively measured. The half life of the stored X-ray image on the IP is about 400 h. Small random dots appear in the image due to a small number of radioactive radium atoms contained in the phosphor. These dots are negligible if the IP is read within a couple of hours after it is kept in the dark or if the image is read out under a relatively low sensitivity of the image reader.

4 Application

4.1 Small Angle Diffraction from Contracting Muscle [19,20]

X-ray diffraction patterns from contracting muscle have been previously studied in time-resolved mode with gas-type one-dimensional detectors [21–26]. These studies have provided information on the molecular changes that occur during force development and sliding movement. Another approach to the time-revolved structure analysis

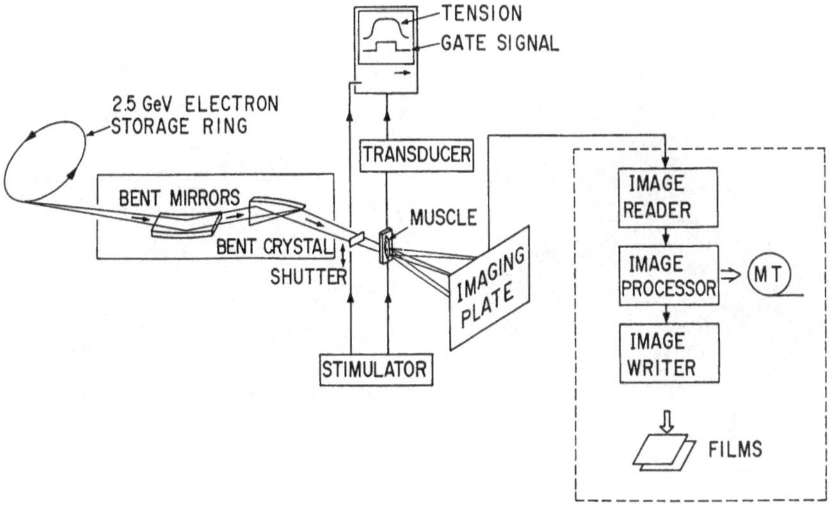

Fig. 10. Experimental configuration for recording two-dimensional X-ray diffraction patterns from contracting muscles with the imaging plate system (FCR 101) and synchrotron radiation. Synchrotron X-ray radiation from the 2.5-GeV electron storage ring of the Photon Factory was vertically focused by seven 20 cm-long mirrors (fused silica) and horizontally by a bent crystal monochromator (germanium (111), asymmetry cut-angle = 8°). X-ray wavelength was 0.155 nm. The IP system consists of an imaging plate, an image reader, an image processor and an image writer. Muscle tension was produced by electrical stimulation and gate signals for an X-ray beam shutter were monitored and recorded with a storage oscilloscope

has been to study the entire two-dimensional diffraction pattern at high spatial resolution with area detectors. Such experiments had been performed by recording on film the patterns from a long series of contractions [27-29]. However, the exposure time required for satisfactory patterns was too long, and physiological fatigue in the contracting muscle could not be avoided. X-ray TV detector [21,30] and gas-type area detectors, such as multiwire proportional chambers (MWPC) [31,32], are being used for this purpose; thus, the required exposure time has become much shorter. These methods, still do not overcome several problems such as described in the Introduction of this article.

The high sensitivity and wide dynamic range of the imaging plate have resulted in a sufficient reduction in the exposure time to make possible the recording of a clear X-ray diffraction pattern, with up to 2.0 nm axial spacing, from a contracting frog skeletal muscle in as little as 10 s with synchrotron radiation. The sartorius muscle of a bullfrog (~6 mm wide and ~1 mm thick) was mounted in a double-focusing camera [33,34] (Fig. 10). The incident flux of X-rays (wavelength, $\lambda = 0.155$ nm) on the specimen was approximately 8×10^{10} photons/s when the Photon Factory storage ring was operated at 2.5 GeV with a beam current of 145 mA. The recording exposure time ranged from 5 s for strong actin layer lines to 10 s for much weaker high-angle actin layer lines of up to ~2.7 nm spacing. This is a great improvement over the 5 to 10 minutes required for recording a visually similar pattern on X-ray film under the same conditions. The muscle was isometrically tetanized at a sarcomere length of about 2.4 μm (nearly the full overlap length of the thin and thick filaments)

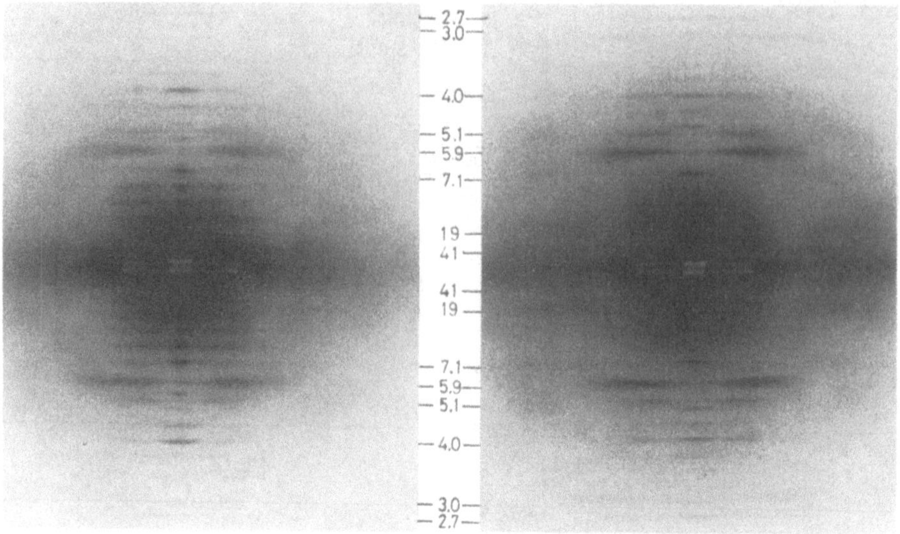

a b

Fig. 11a and b. X-ray diffraction patterns from resting state **a** and isometrically contracting state **b** of a frog skeletal muscle recorded with imaging plates. Exposure time was 10 s. Specimen-to-imaging plate distance was 146 cm. Fiber axis was along the vertical. Numbers are axial spacings (in nm) of the actin layer lines. The intensity ratio of the (1,1) equatorial reflections in the contracting to resting states was about 2.4. This value indicates that the muscle was fully activated in the contracting pattern. Actin layer lines in **b** tended to become slightly sharper in the axial direction. The first layer line observed at 41 nm in **b** is difficult to distinguish from the first myosin layer line remnant in **a**. Three meridional reflections indexed to the first to third orders of a 38.5-nm repeat were present separately from the other actin layer lines, showing a troponin repeat. The intensity on the equator in the central part was attenuated by approximately a factor of 10 with a copper foil 50-μm thick, 3-mm wide and 49-mm long glued to a lead beam stop; the signal was not beyond the dynamic range of the imaging plate. The image reader was preset to give a full range of digital signals (0–255) corresponding to 3.6 orders of magnitude of X-ray intensities; the detective quantum efficiency was maximized

for 1.3 s; this was repeated ten times every 15 s. The diffraction pattern (Fig. 11b) was repeatedly recorded ten times with a 1.0 s exposure time during steady tension. Thus, the total exposure time was 10 s. The tension of the 10th contraction was 83% of the initial tension. The pattern during a resting period (Fig. 11a) was recorded from the same muscle before contraction in the same exposure time. There appeared strong 5.9- and 5.1-nm actin layer lines; except for these two reflections, the actin layer lines are extremely weak. The strong layer-line series with a repeat of 42.9 nm (which appeared like breastbones in the region inside the 5.9-nm actin layer line) are due to the regular helical arrangement of the myosin heads around the thick filament backbone. In contrast, the myosin off-meridional layer lines disappeared and the actin layer lines were distinctly observed during contraction (Fig. 11b).

Changes of actin layer lines during contraction: To show the intensity change of the actin layer lines during contraction, we subtracted the digital data of the resting pattern from that of the contracting pattern (Fig. 12). As can be seen in Fig. 12, the diffraction intensities of most of the actin layer lines (indexed by reciprocal spacings) increased. For instance, the intensity of the 2.7-nm meridional reflection that corres-

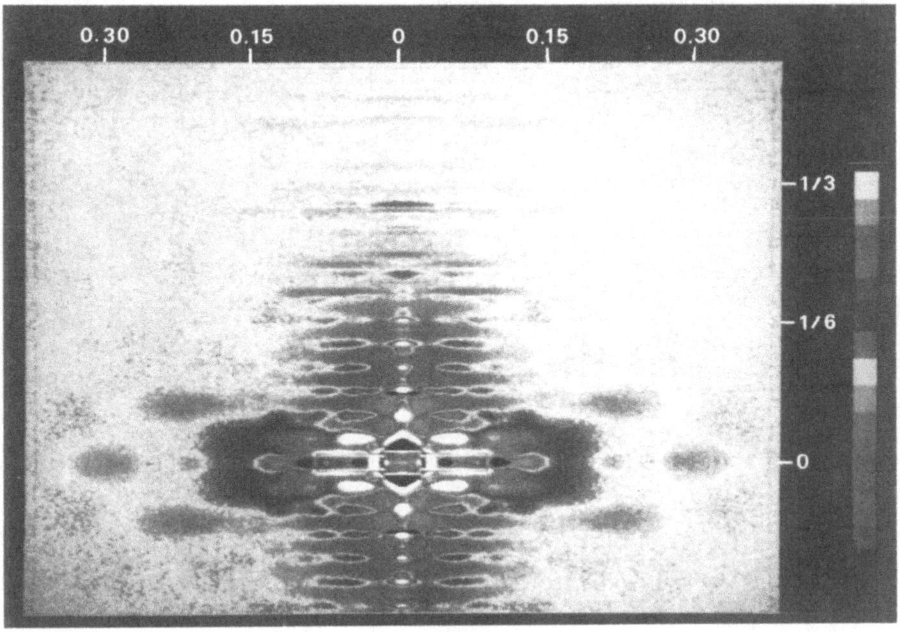

Fig. 12. A difference pattern obtained by subtracting the digital data of the resting pattern (Fig. 11 b) from those of the contracting pattern (Fig. 11 a). Numbers on the top and right-hand side are reciprocal radial and axial coordinates (in nm^{-1}), respectively. The yellow (0 level) through red to dark violet color indicate increasing intensities of layer-line reflections and the light green to white colors indicate the decreasing reflection intensities during contraction. Almost all actin-based layer lines increased in intensity. Intensified broad reflections on the row lines at reciprocal radial spacings around 0.15 to 0.3 nm^{-1} between the first and the 2.7-nm layer lines were 11-, 8.6- and 7.1-nm layer lines, the reflection on the 5.1-nm layer line, 3.9- and 3.0-nm layer-line reflections. The dark violet reflections on the equator and the meridian are the (1,1) and about (4,0) equatorial Bragg reflections and the 14.5-nm myosin meridional reflections, respectively. At radial positions of 0.22 nm^{-1} and 0.32 nm^{-1} on the equator, broad (unsampled) reflections were intensified. The cause of intensification of these equatorial reflections is unknown, but the time course of the intensity change of the reflection at 0.22 nm^{-1} was close to those of the (1,1) equatorial and the 42.9-nm myosin reflections [26]

ponds to the axial repeat of actin monomers in F-actin increased by about 15%. The prominent 5.1- and 5.9-nm reflections increased in the integrated intensity by ~100% and 20 to 50%, respectively. These reflections correspond to the pitches of the two helices of the actin monomers in the filaments. There were other layer-line reflections in several places on the row lines located between 0.15 and 0.3 nm^{-1} from the meridian. The largest intensification of these reflections was observed on the layer line at an axial spacing of ~1/19 nm^{-1}, corresponding to the second actin layer line. The strong appearance of this reflection has been attributed to a structural change of the tropomyosin in the thin filament on activation [26]. The first layer line that corresponds to the crossover repeat of the double strands of the F-actin helix was weak and difficult to distinguish from the remnant of the first myosin layer line. Thus, the change was not clear in the inner side; however, a slight and distinct intensification was noted in the radial range of 0.2 to 0.3 nm^{-1}. The axial spacing of the first layer line was about 41 nm, not 37 nm as observed in the rigor muscle. However, most of

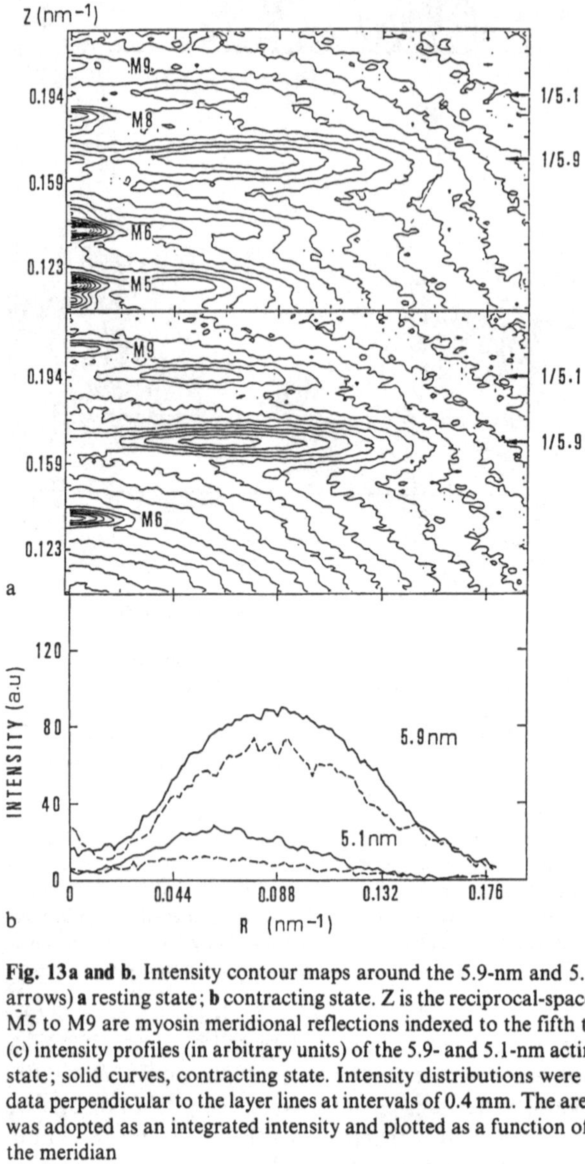

Fig. 13a and b. Intensity contour maps around the 5.9-nm and 5.1-nm actin layer lines (indicated by arrows) **a** resting state; **b** contracting state. Z is the reciprocal-space axial coordinate from the equator. M5 to M9 are myosin meridional reflections indexed to the fifth to ninth orders of a 42.9-nm repeat. (c) intensity profiles (in arbitrary units) of the 5.9- and 5.1-nm actin reflections. Dashed curves, resting state; solid curves, contracting state. Intensity distributions were measured by scanning the intensity data perpendicular to the layer lines at intervals of 0.4 mm. The area of the peak above the background was adopted as an integrated intensity and plotted as a function of the reciprocal coordinate (R) from the meridian

the actin layer lines did not appreciably change shape or peak positions in the radial direction, nor were their axial spacings significantly different from those in the resting state. As an example, a comparison of the intensity contour maps around the 5.9- and 5.1-nm layer lines is shown in Fig. 13 together with the intensity profiles of the 5.9- and 5.1-nm reflections along the layer lines in the resting and contracting patterns. These two layer lines became slightly sharper in the axial direction but did not shift their peak positions in the radial and axial directions. Thus, the myosin heads seem to geometrically mark the actin filaments during contraction.

In a rigor muscle, the actin layer-line intensities seemed, in general, to be greater than in the contracting muscle and most of the peaks shifted towards the meridian. The whole pattern had a ladder-like appearance with a strong first layer line at $\sim 1/37\ nm^{-1}$. This pattern has been shown to arise from a periodic attachment of myosin heads to actin with the symmetry of the actin filaments [27]. Thus the time-averaged structure of the actin filament during steady isometric tension was distinctly different from that of the actin filament in rigor.

The intensification of the actin layer lines remained for the stretched muscle with a small overlap of the thin and thick filaments. This intensity change seemed to be larger than expected, given that the change occurs proportionally to the number of myosin heads that were present in the overlap zone. Recent time-resolved studies have shown that the intensity changes of the 5.9 nm and second layer-line reflection preceded those of the myosin and equatorial reflections on activation [23,26]. Taken together, there could also be a change in the structure of the actin monomer itself that takes place in the process of activation by an interaction with the myosin heads, as well as by the structural changes of regulatory proteins. This might contribute to the intensity enhancement of the actin layer lines during contraction.

Changes of the myosin layer lines during contraction: Off-meridional parts of the myosin layer lines and many so-called forbidden meridional reflections decreased greatly in intensity or disappeared. However, the 42.9 nm first off-medidional reflections seemed to remain within about 20% of the resting intensity. The meridional reflections indexed to 3n orders of 42.9 nm repeat were strong up to the 24th one, with a 1 to 3% increase in the axial spacing. The width of these reflections perpendicular to the meridian approximately doubled. If the widths were taken into account, the intensities of most of these reflections even increased, indicating a strengthening of the threefold screw symmetry of the myosin filament and a better ordering of myosin projections in planes with an axial spacing of 14.5 nm. In a rigor muscle, these meridional reflections could still be observed, but were weakened; the 42.9-nm layer line completely disapperared. In agreement with observations from electron microscopic studies of a rapidly frozen sample [35], our results suggest that, during contraction, the myosin heads retain the axial periodicity of the myosin filament and align more perpendicularly along the actin filament.

4.2 Protein Laue Diffraction of Cytochrome c′ [36]

Utilizing a white synchrotron radiation beam, Laue diffraction patterns from cytochrome c′ were recorded (Fig. 14). The crystal ($0.05 \times 0.07 \times 0.2\ mm^3$) was exposed to the white beam (0.1 ~ 0.19 nm) which had been transmitted through an Al filter and reflected by a plane mirror of fused quartz. The patterns were obtained with an X-ray dose of approximately one thirtieth as much as that required by a high-sensitivity X-ray film. More than 50 exposures could be obtained from one crystal specimen without any noticeable radiation damage. On the contrary, when X-ray film is used, a decrease in the intensities of reflections was observed after two exposures, showing that the crystal began to be damaged by X-ray irradiation. Thus, the most advantageous point of the IP in the Laue diffraction experiment is the high detective quantum efficiency which results in a great reduction of the required X-ray dose. The other

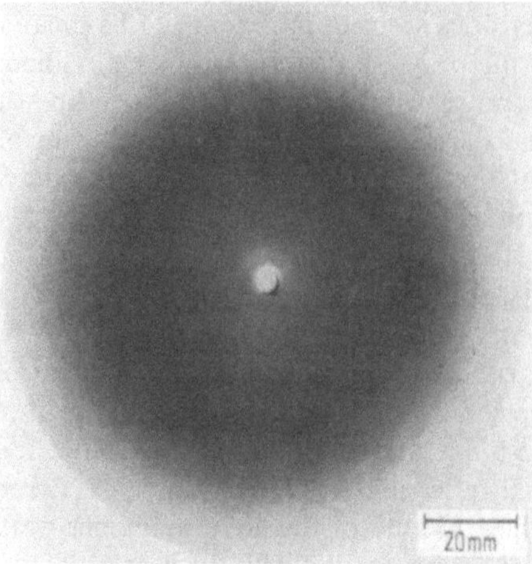

Fig. 14. A Laue diffraction photograph from cytochrome c′ using quasi-white X-ray radiation (λ: 0.1–0.19 nm)

advantages of the IP are that i) the response to X-rays of different wavelengths is relatively uniform, because the absorption efficiency of the IP is almost 100% for X-rays of wavelength longer than 0.07 nm and that ii) due to the wide dynamic range of the IP, the use of multiple film technique is not required to cover the full range of intensities; thus, a correction for intensity analysis becomes much simpler. Recently, a Laue diffraction pattern from lysozyme crystal was recorded with an exposure of 800 μs using a focused X-ray beam from a normal bending magnet at the Photon Factory[37].

If the experiment is carried out using more intense synchrotron radiation from

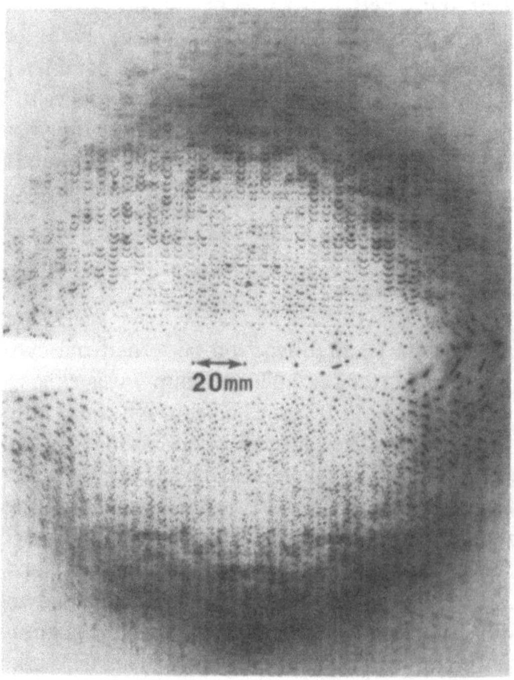

Fig. 15. A Weissenberg photograph of ω-amino acid:pyruvate aminotransferase (mercury derivative). A single bent triangular Si(111) monochromator and a Weissenberg camera for macromolecular crystallography were used. The oscillation angle: 30°; the radius of the camera: 287 mm, exposure time: 4.7 min

multi-pole wigglers, the exposure time will be reduced to the region of a few tens of microseconds. Such a reduction of the exposure time by the use of IPs will open a new phase in time-resolved Laue diffraction experiments of protein crystals.

4.3 Weissenberg Photograph of ω-amino Acid: Pyruvate Aminotransferase [16]

Figure 15 shows a Weissenberg photograph of ω-amino acid: pyruvate amino transferase (space group: I 222, cell dimensions: a = 12.46, b = 13.79, c = 6.145 nm)[38]. The photograph was obtained with a Weissenberg camera [39] designed for macromolecular crystallography. A full data set was collected from eight such patterns with the computer program WEIS [40] and the R_{sym} value for the intensities was 5.0%. The exposure time for each pattern was 4.7 min. By using the IP system, the phase determination of this protein could be performed by the anomalous dispersion method with the multi-wavelength data sets collected from only one crystal.

4.4 Time-Resolved SAXS from a Polymer Sheet During Stretch [41]

The IP system seems apparently to be poorly suited to time-resolved measurements when compared with electronic area detectors such as MWPCs which permit a time-

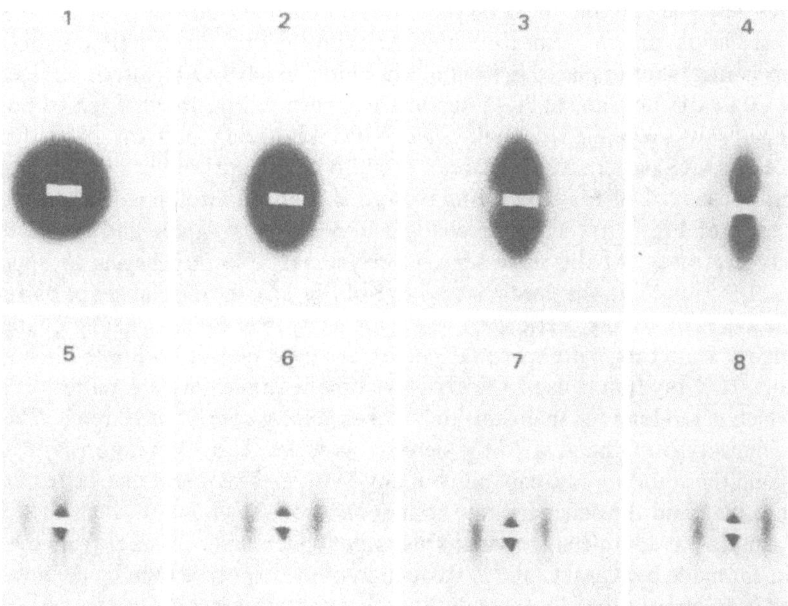

Fig. 16. Time-resolved small-angle X-ray scattering patterns from polyethylene sheet recorded during stretch (12 mm/min, 32% stretch/min) in the horizontal direction. An exposure time for each pattern was 1 s. Intervals between exposures were 70 s. An X-ray wavelength was 0.155 nm

resolution of a millisecond or better. However, the IP can be used and is of course not counting rate limited like MWPC's for time-resolved measurements using synchrotron radiation. When an IP is used with a fast mechanical shutter the X-ray beam can be gated with a time-resolution of a millisecond. Hence, an X-ray pattern during a particular phase of a transient phenomenon can be recorded on the IP with a time-resolution of a millisecond. X-ray patterns during different phases of the phenomenon can also be recorded on different IPs by changing the timing of the gate signals of the shutter. In this method, the experiment has to be repeated at least n times when the number of time-slices is n. The total number of repetitions is, however, smaller with IPs than with a MWPC. This is because of the lack of counting rate limitations whilst maintaining a high DQE.

The other way to perform time-resolved measurements using IPs is to exchange IPs rapidly and successively in combination with a mechanical shutter. This method is valid for the measurements in which a time-resolution is of the order of a sub-second or longer.

Figure 16 shows an example of this method applied to study small-angle X-ray scattering (SAXS) from a polyethylene sheet under stretch. The polyethylene sheet ($\varrho \sim 0.918\,g/cm^3$, melt flow rate (MFR) $\sim 0.9\,g/10\,min$) of 2 mm thickness was placed in the doubly focusing camera [33, 34] and stretched at a speed of 12 mm/min (32% stretch/min) in the horizontal direction by a DC motor-driven stretching device. Both ends of the sample were stretched with equal velocities in the opposite directions so that the center of the X-ray beam can impinge on an identical position of the sample during stretch. Eight imaging plates ($126 \times 126\,mm^2$) were held in separate film cassettes, and they are mounted on a remotely controlled turn-table which is a part of the oscillation camera of Enraf-Nonius. An exposure time for each IP was controlled by a mechanical shutter placed in front of the sample. Each SAXS pattern was recorded with a 1.0 s exposure time at 70 s intervals As is seen in Fig. 16, an oriented polymer no longer shows symmetric small-angle X-ray scattering patterns. The dynamic changes of SAXS patterns during stretch seem to be categorized into three phases. In the first phase (1, 2 of Fig. 16), a ring pattern is changed into an ellipse. In the next phase (3, 4 of Fig. 16), the ellipse changes into a dipolar shape and the scattering intensity decreases. At the same time, a meridional reflection begins to appear at $1/9.0 \sim 1/9.5\,nm^{-1}$. In the third phase (5 to 8 of Fig. 16), the dipolar shape disappears and the intensity of the meridional reflection increases. Such dynamic changes of asymmetric scattering patterns could not be observed clearly with one-dimensional detectors. If X-ray film is used, the exposure time required for each pattern is about 40 s, which is too long for snapshots to be taken during continuous stretch. The total X-ray intensity over the area of the detector is more than 10^7 photons/s, which is far beyond the counting-rate capability of any MWPC. The meridional reflection was split into two, and thus changed into four-spots (not shown here), when the tension of the sample was completely relaxed. This agrees with results obtained from the static experiment made by Kasai et al. [42]. Reduction of the exposure time by the use of IPs enabled us to observe the process under stretch separately from the relaxation process. Detailed analysis of these SAXS patterns will appear elsewhere [43].

In order to achieve a sub-second time-resolution using IPs, a rapid IP-exchanger has been designed in combination with a rapid mechanical shutter [44]. Figure 17 shows a rear view of the IP exchanger. The principle of the IP exchanger is similar

Fig. 17. A rear view of the rapid exchanger for imaging plates. The size of the IPs is 126×126 mm^2 ($5'' \times 5''$). A minimum exposure time for each IP is designed to be 0.1 s and a dead time between exposures is 0.2 s. Thus, the system allows 3.3 exposures per second. The controller (not shown here) synchronizes the motion of IPs with that of the mechanical shutter (not shown here) placed in front of specimens

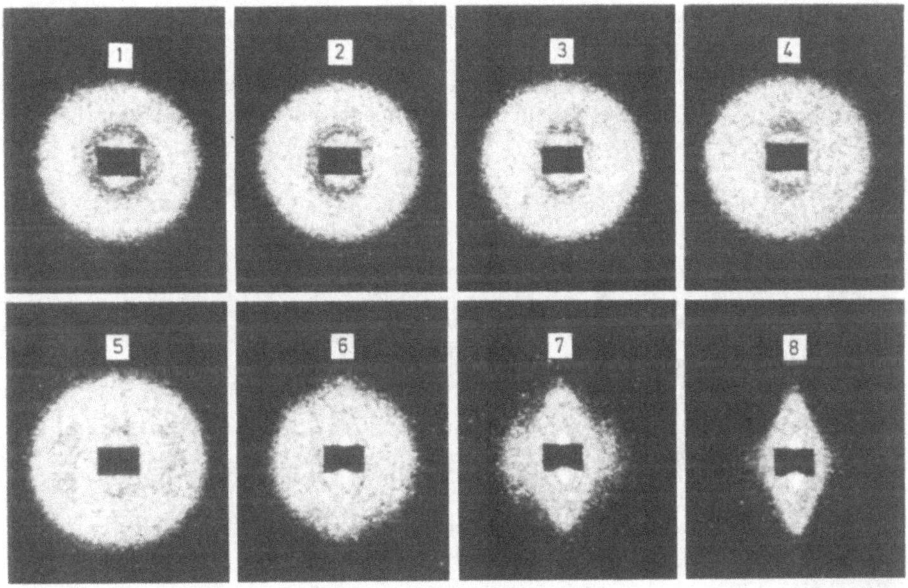

Fig. 18. Time-resolved small-angle X-ray scattering patterns from polypropylene sheet under quick stretch in the horizontal direction. A speed of stretch was 233 mm/min (367 % stretch/min). An exposure time for each pattern was 0.1 s. Intervals between exposures were 0.2 s. An X-ray wavelength was 0.155 nm. A slight deformation of the symmetric SAXS pattern was already observed in the second patterns, suggesting some degree of orientation was brought about in quite an early stage. The SAXS patterns changed abruptly and drastically in the sixth pattern just when the sample began to yield (when the tension began to decrease).

to that of the vertical card reader of computers. IPs are mounted separately in IP cassettes, and they are placed vertically in the left-side bin to make a line. The foremost IP in the bin is pushed rightward to the position for exposure (middle of the system) by a motor-driven cam system. At this moment, the remaining IPs in the bin are pushed forward by a spring behind them and the second IP proceeds foremost. The IP at the position for exposure is again pushed rightward to the right-side bin by the second IP cassette when the second one is likewise pushed to the position for exposure. A size of the IPs is 126×126 mm^2 (5″ × 5″). The minimum exposure time is 0.1 s and the dead time between exposures is 0.2 s; thus, the system permits 3.3 exposures per second. Both the exposure time for each IP and the interval time between two exposures are adjustable with an arbitrary increment in units of 0.1 s. The number of IPs which can be mounted in this system is forty. The reproducibility of the IP position for exposure is within ± 0.05 mm in two orthogonal directions. These specifications allow multiple exposures to be obtained with good photon statistics in time-resolved X-ray patterns. Figure 18 shows a color graphic display of time-resolved small-angle X-ray scattering patterns from a polypropylene sheet during stretch obtained by using this IP-exchanger system. An exposure time for each pattern was 0.1 s and intervals between two exposures were 0.2 s. The sample was stretched in a speed of 233 mm/min (367% stretch/min) at the ends of the sample in the horizontal direction. A pair of notches was made in advance on the sample so that the deformation should occur at the position where the X-ray beam impinged. When the stretch velocity was made slower to 1.8 mm/min (2.8% stretch/min), each SAXS pattern differed considerably (not shown here) from the one observed at a corresponding stretch ratio under faster stretch velocity. Detailed results of this experiment will be discussed elsewhere [43)].

Fig. 19. A front view of the drum-type image reader and image writer. An imaging plate is attached on the drum (right-hand side) with adhesive tapes. In the black container behind the drum, a He-Ne laser and a laser-focusing optics are accommodated together with PMTs which collect the photo-stimulated luminescence. The black container moves in the rightward direction along the drum axis, while the drum rotates; thus, the focused He-Ne laser beam scans the surface of the IP on the drum. The circumference and length of the drum are 400 mm, 300 mm, respectively. The rotation speed of the drum can be at 2.5, 5.0 and 10.0 rotations per second. A graphic film is attached on the left-hand side drum so that the digital image is imprinted on it by modifying the intensity of a glow lump

5 Development of a new Image Reader [45)]

As the original IP system was designed for diagnostic radiography, there is still some room left for improvements for the purpose of diffraction studies. A new image reader has been built at the Photon Factory by modifying a conventional drum-type film densitometer (Fig. 19). In the new image reader, the following points have been improved: i) Detective quantum efficiency is further increased by the use of a glass device which efficiently collects the PSL (Fig. 20). Moreover, the most powerful He—Ne laser beam possible is used together with more efficient laser optics in order to increase the efficiency to stimulate the PSL. ii) The pixel sizes of $25 \times 25\ \mu m^2$ and $50 \times 50\ \mu m^2$ in addition to $100 \times 100\ \mu m^2$ are available by making both a focus size and a scanning pitch of the laser beam changeable. The thickness of the phosphor layer of the IP should be decreased so as to match the spatial resolution expected with smaller pixel sizes. iii) A 12-bit A/D converter is used in place of a 8-bit A/D converter to improve the precision in X-ray intensity measurement. The uncertainty in the intensity measurement due to the digitization process is reduced from 2.6 to 0.2%, when the system is preset to have a full range of the A/D converter corresponding to 3 orders of magnitude (the dynamic range of a single PMT). iv) In order to fully utilize the dynamic range of the IP, the entire intensity range of the PSL radiation

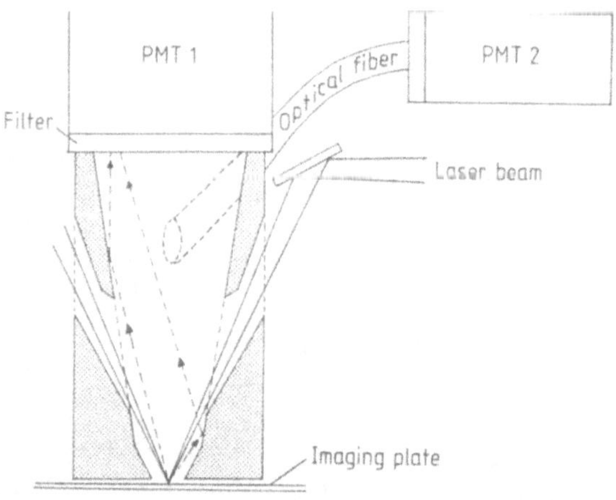

Fig. 20. A schematic drawing to show the way in which the imaging plate is stimulated by a laser beam and the photostimulated luminescence (PSL) is collected by the two photomultiplier tubes (PMTs). A dotted area illustrates a device made of glass which was designed to gather the PSL as efficiently as possible. The PSL from the IP is reflected on the inner-surfaces of the glass which were multi-coated to increase the reflectivity for the PSL ($\lambda = 390$ nm) and to decrease the reflectivity for the scattered laser beam ($\lambda = 633$ nm). The device consists of two octagonal pyramids to form the shape of a paraboloid. The device is placed 1 mm apart from the surface of the IP so that it covers approximately 80% of the total solid-angle centered upon the spot where the PSL is emitted. The device has three holes; two holes are used to let laser beams pass through and the third one is used to guide 1% of the PSL into the PMT2 through an optical fiber. The blue filters at the entrances of the PMTs shield stray and scattered laser light completely to reduce the background level of the system.

is detected simultaneously by two PMTs each with different sensitivities. The second PMT (PMT2), receiving roughly 1 % of the total amount of the PSL, is used to detect the PSL at a higher intensity range where the first one (PMT1) is saturated. v) Arbitrary sizes of the IP can be read. vi) In order to decrease the image distortion, a drum type film densitometer is utilized which provides more accurate scanning pitches than the flat one. vii) The non-uniformity of response is improved. This is because in the drum-type image reader the incident angle of the laser beam against the IP surface is always constant; thus, the laser beam can stimulate the phosphor more uniformly everywhere on the surface of the IP. viii) A high-resolution graphic display system (1280 × 1024 pixels, 8-bit/pixel, 19″) is also available in addition to the image writer. The graphics system saves the running cost of film for display as well as the time required for judging the experimental results.

A block diagram of the system is shown in Fig. 21. The IP exposed to the X-ray pattern is attached on the drum of the image reader using adhesive tapes under a normal safety light. The system is controlled by a 32-bit minicomputer. The output signals from the PMTs are digitized by 100 kHz A/D converters and the digital data are stored in disk. It takes 200 s to read the image on the IP of 250×200 mm^2 with a pixel size of 100×100 μm^2. The digital image is displayed on the color graphic display immediately after the readout. The 12-bit digital data are modified into 8-bit data on the graphic display according to the threshold level and dynamic range which are presetable from the keyboard terminal.

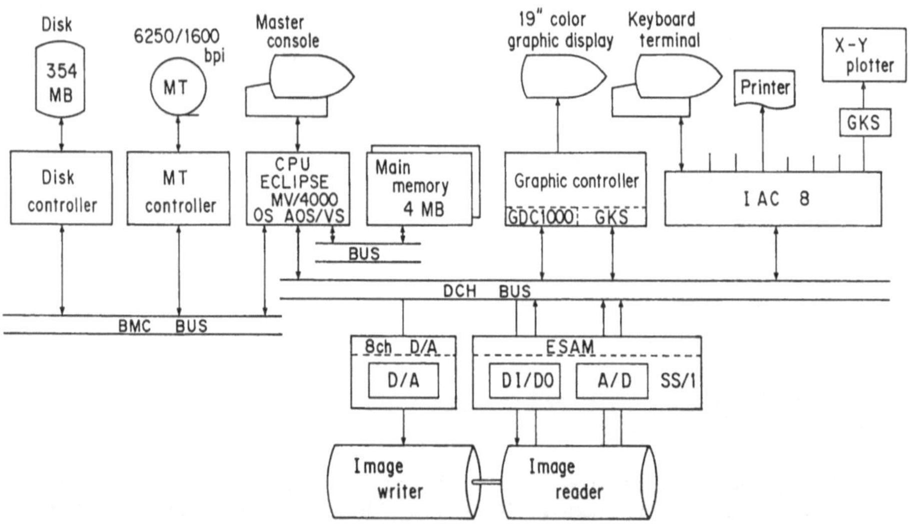

Fig. 21. Block diagram of the image reader and image writer using the drum-type film scanner. The image data are digitized by 100 kHz A/D converters (12-bit) and are stored in winchester disk. The digital image data are immediately displayed on the 19″ color graphic display. The data are in the end dumped on magnetic tape (6250/1600 bpi). The parameters which can be preset for image reading are, 1) the size of IP, 2) the pixel size, 3) the detective quantum efficiency (high voltages for PMTs), 4) the rotation speed of the drum and 5) whether one or two PMTs are used

6 Conclusion

The excellent characteristics of the IP system as an integrating area detector are well suited to X-ray diffraction and scattering experiments using synchrotron radiation. The system is particularly well suited for biological specimens which require the shortest exposure time or the smallest amount of X-ray dose possible as demonstrated through the applications in this article.

Only two years have passed since 1985 when we, for the first time, demonstrated the quantitative performance and the validity of the IP system for X-ray diffraction studies using synchrotron radiation. Besides the applications which are described here, we believe that the IP system will be very useful in such experiments as X-ray diffraction under high pressure and high temperature, Compton scattering, X-ray diffuse scattering, dispersive EXAFS, X-ray topography and X-ray microscopy. The IP system will partially replace conventional X-ray film, and even multiwire proportional chambers, in many application fields of synchrotron radiation. When more intense X-rays are available for example from insertion devices installed in a $6 \sim 8$ GeV storage ring, the importance of the IP system will become even greater because of its high quantum efficiency and the lack of any counting rate limitation.

7 Acknowledgement

We would like to thank our collaborators who appeared in the original papers referred to in each section. Special thanks are due to Dr. N. Kamiya (The Institute of Physical and Chemical Research) who worked together to evaluate the IP system and Mr. T. Kojima and H. Nomura (Sumitomo Chemical Co. Ltd.) who worked together in the time-resolved SAXS experiments for polymers. We are grateful to Drs. N. Kamiya, N. Sakabe who took the photograph of Fig. 15 and to Dr. H. Tanaka who helped us in the experiment on muscle. We would like to appreciate the support and advice from Prof. M. Ando (PF) for constructing the rapid IP exchanger. The construction of the image reader and the rapid IP exchanger was supported by a grant-in aid for developmental scientific research (No. 60850013) and for scientific research (No. 62420053), respectively, from the Ministry of Education, Science and Culture, Japan.

8 References

1. Cork, C., Fehr, D., Hamlin, R., Vernon, W., Yuong, N. H., Perez-Mendez, V.: J. Appl. Cryst. *7*, 319 (1973)
2. Hendricks, R. W.: J. Appl. Crys. *11*, 15 (1978)
3. Gabriel, A., Dauvergne, F., Rosenbaum, G.: Nucl. Instrum. Methods *152*, 191 (1978)
4. Kahn, R., Fourme, R., Caudron, B., Bosshard, R., Benoit, R., Bouclier, R., Charpak, G., Santiard, J. C., Sauli, F.: ibid. *172*, 337 (1980)
5. Phizackerley, R. P., Cork, C. W., Hamlin, R. C., Nielsen, C. P., Vernon, W., Xuong, N. H., Perez-Mendez, V.: ibid. *172*, 393 (1980)
6. Boulin, C., Dainton, D., Dorrington, E., Elsner, G., Gabriel, A., Bordas, J., Koch, M. H. J.: ibid. *201*, 209 (1982)
7. Bade, D., Parak, F., Mössbauer, R. L., Hoppe, W., Levai, N., Charpak, G.: ibid. *201*, 193 (1982)

8. Helliwell, J. R., Hughes, G., Przybylski, M. M., Ridley, P. A., Sumner, I., Bateman, J. E., Connolly, J. F., Stephenson, R.: ibid. *201*, 175 (1982)
9. Reynolds, G. T., Milch, J. R., Gruner, S. M.: Rev. Sci. Instr. *49*, 1241 (1978)
10. Arndt, U. W., Gilmore, D. J.: J. Appl. Cryst. *12*, 1 (1979)
11. Schulz, G. E., Rosenbaum, G.: Nucl. Instrum. Methods *152*, 205 (1978)
12. Sonoda, M., Takano, M., Miyahara, J., Kato, H.: Radiology *148*, 833 (1983)
13. Kato, H., Miyahara, J., Takano, M.: Neurosurg. Rev. *8*, 53 (1985)
14. Takahashi, K., Kohda, K., Miyahara, J., Kanemitsu, Y., Amitani, K., Shionoya, S.: J. Lumin. *31* & *32*, 266 (1984)
15. Takahashi, K., Miyahara, J., Shibahara, Y.: J. Electrochem. Soc. *132*, 1492 (1985)
16. Miyahara, J., Takahashi, K., Amemiya, Y., Kamiya, N., Satow, Y.: Nucl. Instrum. Methods *A246*, 572 (1986)
17. Amemiya, Y., Kamiya, N., Satow, Y., Matsusita, T., Wakabayashi, K., Tanaka, H., Miyahara, J.: in Biophysics and Synchrotron Radiation (Castellano, A. C. and Invidia, L. Eds.) Springer-Verlag, Berlin 1987, pp. 61–72
18. Dainty, J. C., Shaw, R.: Image Science (Academic Press, New York, 1974)
19. Wakabayashi, K., Amemiya, Y., Tanaka, H.: Proc. 11th Taniguchi Int. Symp. Biophys. (ed. Yanagida, T.) Taniguchi Foundation, 1986, pp. 56–62
20. Amemiya, Y., Wakabayashi, K., Tanaka, H., Ueno, Y., Miyahara, J.: Science *237*, 164 (1987)
21. Huxley, H. E., Faruqi, A. R., Koch, M. H. J., Milch, A. R.: Nature *284*, 140 (1980)
22. Huxley, H. E., et al.: J. Mol. Biol. *169*, 469 (1983)
23. Wakabayashi, K., Tanaka, H., Amemiya, Y., Fujishima, A., Kobayashi, T., Hamanaka, T., Sugi, H., Mitsui, T.: Biophys. J. *47*, 847 (1985)
24. Wakabayashi, K., Tanaka, H., Kobayashi, T., Amemiya, Y., Hamanaka, T., Nishizawa, S., Sugi, H., Mitsui, T.: ibid. *49*, 581 (1986)
25. Tanaka, H., Kobayashi, T., Amemiya, Y., Wakabayashi, K.: Biophys. Chem. *25*, 161 (1986)
26. Kress, M., Huxley, H. E., Faruqi, A. R., Hendrix, J.: J. Mol. Biol. *188*, 325 (1986)
27. Huxley, H. E., Brown, W.: ibid. *30*, 383 (1967)
28. Elliott, G. F., Lowy, J., Millman, B. M.: ibid. *25*, 31 (1967)
29. Haselgrove, J. C.: ibid. *92*, 113 (1975)
30. Tregear, R. T., Milch, J. R., Goody, R. S., Holmes, K. C., Rodger, C. D.: in Cross-Bridge Mechanism in Muscle Contraction (Sugi, H., Pollack, G. H., Eds.) Univ. Tokyo Press, of Tokyo 1979, pp. 407–423
31. Matsubara, I., Yagi, N., Miura, H., Ozeki, M., Izumi, T.: Nature *312*, 471 (1984)
32. Huxley, H. E., et al.: in Maeda, Y. in Structural Biological Applications of X-ray Absorption, Scattering and Diffraction (Bartunik, H. D., Chance, B., Eds.) Academic Press, New York 1986, pp. 407–423
33. Amemiya, Y., Wakabayashi, K., Hamanaka, T., Wakabayashi, T., Matsushita, T., Hashizume, H.: Nucl. Instrum. Methods *208*, 471 (1983)
34. Wakabayashi, K., Hamanaka, T., Hashizume, H., Wakabayashi, T., Amemiya, Y., Matsushita, T.: in X-ray Instrumentation for the Photon Factory: Dynamic Analyses of Micro Structures in Matter, Hosoya, S., Iitaka, Y., Hashizume, H., Eds.) KTK Science, Tokyo 1986, pp. 61–74
35. Tsukita, S., Yano, M.: Nature *317*, 182 (1985)
36. Satow, Y., Amemiya, Y., Matsushita, T.: Photon Factory Activity Report *4*, 326 (1986)
37. Satow, Y., Amemiya, Y., Matsushita, T., Phizackevley, R. P.: Photon Factory Activity Report *5*, 158 (1987)
38. Morita, Y., Aibara, S., Yonaha, K., Toyama, S., Soda, K.: J. Mol. Biol. *130*, 211 (1979)
39. Sakabe, N.: J. Appl. Cryst. *16*, 542 (1983)
40. Higashi, T., Kamiya, N., Matsushima, M., Sakabe, K., Sakabe, N.: Photon Factory Activity Report (1983/84) p. VI–16
41. Amemiya, Y., Kojima, T., Nomura, H., Chikaishi, K., Satow, Y., Matsushita, T.: Photon Factory Activity Report *4*, 306 (1986)
42. Kasai, N., Kakudo, M.: J. Polymer Science *A 2*, 1955 (1964)
43. Amemiya, Y., et al.: manuscript in preparation
44. Amemiya, Y., Ando, M. et al.: manuscript in preparation
45. Amemiya, Y., Matsushita, T., Nakagawa, A., Satow, Y., Miyahara, J., Chikawa, J.: Nucl. Instrum. Methods *A266*, 645 (1988)

Photoacoustic X-ray Absorption Spectroscopy

Tsutomu Masujima

Institute of Pharmaceutical Sc ences, Hiroshima University School of Medicine, Kasumi 1-2-3, Hiroshima 734, Japan

Table of Contents

The transient heat generation of material due to X-ray absorption was found to be detected as the photoacoustic effect. The photoacoustic signal which was detected by a microphone linearly corresponded to the ring current of the storage ring and to the ion chamber current for monochromatic X-ray beam, and showed the maximum intensity for metal foil when the thickness of the foil was close to the absorption length. The spectroscopic study of this heat generation effect revealed not only the inner shell edge absorption but also the fine structure at the edge absorption region. The spectrum for polycrystal copper quite corresponded to the extended X-ray absorption fine structure (EXAFS). The Fourier transform analysis of the spectrum gave the atomic distances of copper atoms. On the basis of this results, depth profiling of the nickel plated copper by EXAFS and imaging analysis of model sample with focused X-ray beam scanning were performed. The future possibility of this method which extended the photoacoustic spectroscopy into the X-ray region was briefly discussed.

1 Introduction

When materials are irradiated by X-rays, several effects occur, e.g. absorption, scattering, and X-ray fluorescence. The heat should be generated by the absorption of electro-magnetic wave even by X-ray as shown in Fig. 1. The heat generation was actually observed in, for example, the deterioration of a silicon single crystal by strong X-ray. However, so much attention has not been paid in scientific sense to this phenomena.

The photoacoustic method, especially microphonic detection is one of the most sensitive method to detect the heat generation from the solid surface [1]. Figure 2 shows the principle of the photoacoustic method which is originated from the idea of A. G. Bell 100 years ago [2]. When the intermittent light, in this case X-ray, is irradiated to the sample in closed chamber, the heat should be generated by the absorption of photon energy through the energy conversion processes. The heat expands the surrounding gas periodically, and the pressure wave will be generated which can be detected by a microphone. In order to detect 10^{10} cps photon with 9 keV energy

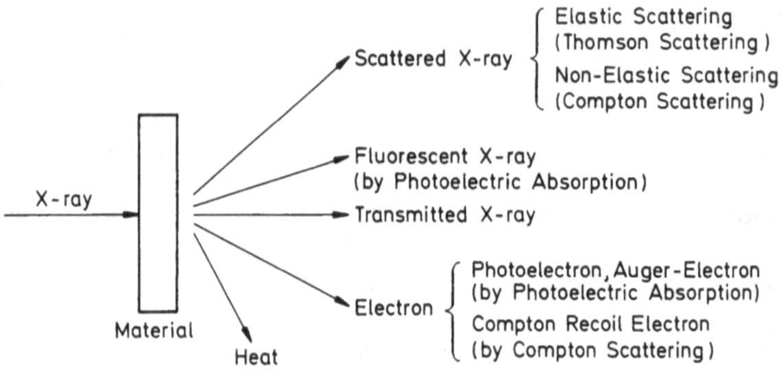

Fig. 1. General aspects of X-ray irradiation after-effects

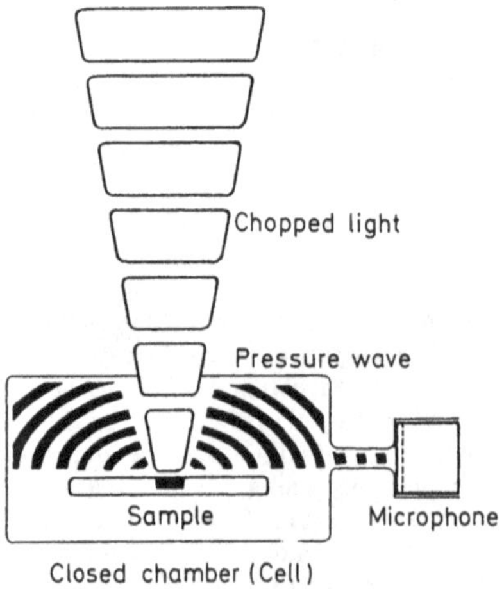

Fig. 2. Principle of the photoacoustic method

at 10 msec time resolution, 1.4×10^{-7} J must be detected. When the heat capacity of the absorber is 2.7×10^{-3} J/K (e.g. Cu 10 μm thick, 10 mm Φ, the temperature increase of 5×10^{-5} deg can be observed. Using the strong X-ray from the electron storage ring at the Photon Factory (PF) of the National Laboratory for High Energy Physics (KEK), Japan, the heat generation by monochromatic X-ray was found to be detected as the photoacoustic effect [3,4], and it was found recently that the heat thus generated reflects atomic informations [5].

The principle of photoacoustic method promises unique potentials as a new tool of analytical method. 1) Samples can be analyzed as it is and 2) non destructive analysis of sub-surface can be done by detecting the time delay for thermal propagation from the subsurface. This preview reports the recent photoacoustic studies in X-ray region which were started from our finding of X-ray photoacoustic effect [3].

2 Instrumentation [3,6]

Figure 3 shows one of our photoacoustic cell for X-ray spectroscopy of solid samples [3]. The cylindrical cell has a sample chamber at the center with volume of 0.16 cm^3 which has two windows of beryllium (18 mm Φ × 0.5 mm thickness). A microphone cartridge is commercially available electret type (10 mm Φ) and the electronics of preamplifier for this microphone is detailed elsewhere [7]. Figure 4 shows the typical experimental setup for spectroscopic study [6]. X-ray was monochromated by channel cut silicon double crystal (111) and ion chamber was set to monitor the beam intensity. Photoacoustic signal intensity was always divided by the ion chamber current for the normalization against the photon flux. X-ray was modulated by a rotating lead plate (1 mm thick) chopper with two blades.

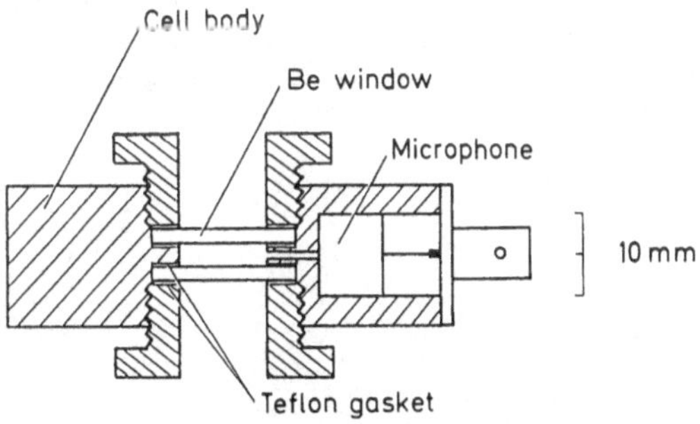

Fig. 3. Cross-section view of photoacoustic X-ray absorption cell

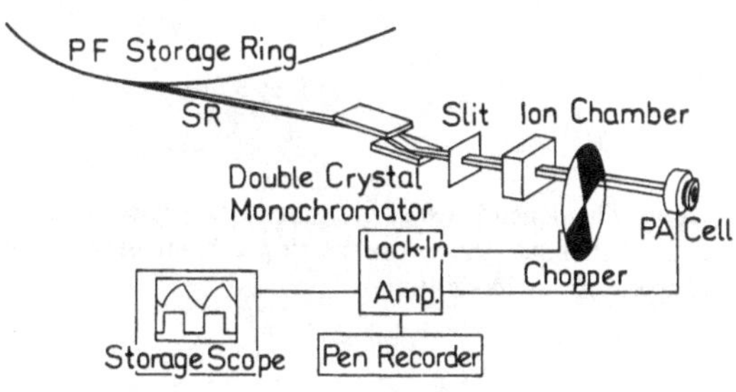

Fig. 4. Schematic setup for photoacoustic X-ray absorption spectroscopy

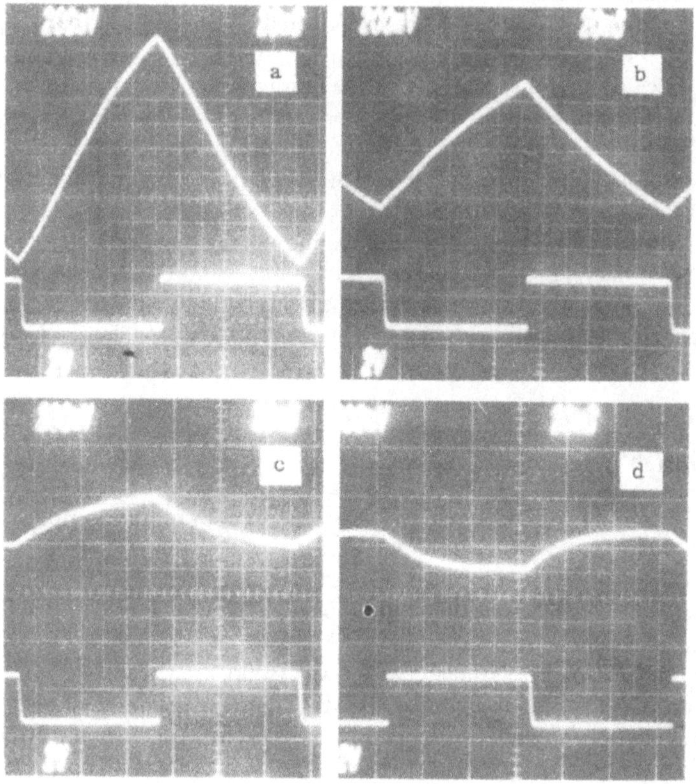

Fig. 5a–d. Photoacoustic signals of solid samples with white X-ray. **a** copper (50 μm thick), ring current (R.C.) 107.4 mA; **b** lead (400 μm), R.C. 140.9 mA; **c** paper (100 μm), R.C. 84.9 mA; **d** no sample, R.C. 87.2 mA. The numerals in figures are scales per division. The pulse signal in each figure is chopping signal. In this case, its phase was inverted by 180 degree due to the setting position of chopper. Thus the lower side of pulse is X-ray on

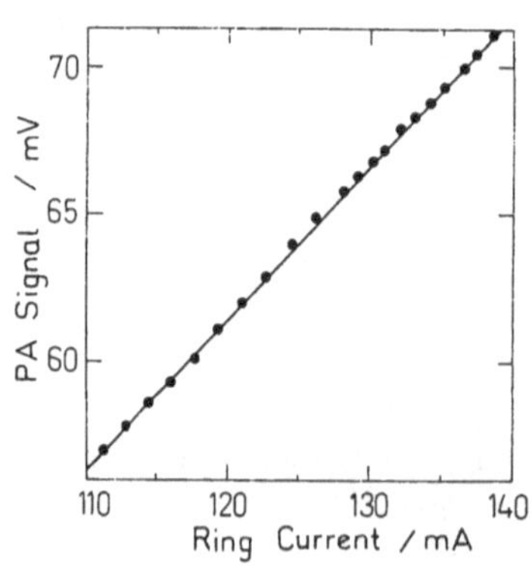

Fig. 6. The dependence of photoacoustic signal intensity of lead (0.4 mm thick) on the ring current in the case of white X-ray irradiation

3 X-ray Photoacoustic Effect of Solid Materials [3,4]

Figure 5 shows the photoacoustic signal of various solid materials for white X-ray. Strong signal for copper (50 μm) (a) shows a typical saw shape with a little convex form rising as seen in the photoacoustic signal by the UV or visible light source. As shown in (d), the signal of the air in the cell chamber (and from the inner faces of beryllium windows) shows a convex shape rising up and seems to overlap on the genuine signals of metal samples as seen in (a) and (b). Thus the original signals of copper, aluminum, and lead should be more triangular shaped. Furthermore, even for the organic material like paper, an weak heat generation was detected as seen in Fig. 5(c). The photoacoustic signal intensity (sample: lead plate (400 μm thick)) for white X-ray was proportional to the ring current in the range of 110–140 mA as shown in Fig. 6. Since the ring current is the measure of the beam photon flux, this method can be applied to the estimation of X-ray dose [8].

Figure 7 is the wave forms of various samples by monochromatic focused X-ray (beam size 1 mm × 1.5 mm) (1.56 Å). The signal amplitude was decreased to 1/100 to 1/200 (by signal intensity/ring current) of those for white X-ray. The detectability of this method was about 10^{10} photon/sec (using copper (50 μm) and at S/N = 1) at this stage. Since the signal of the air (and beryllium windows) was so weak in this case, the triangle shape of the signals for metal samples became clear which shows the rising of temperature at the solid surface. The precise wave-form simulation is under

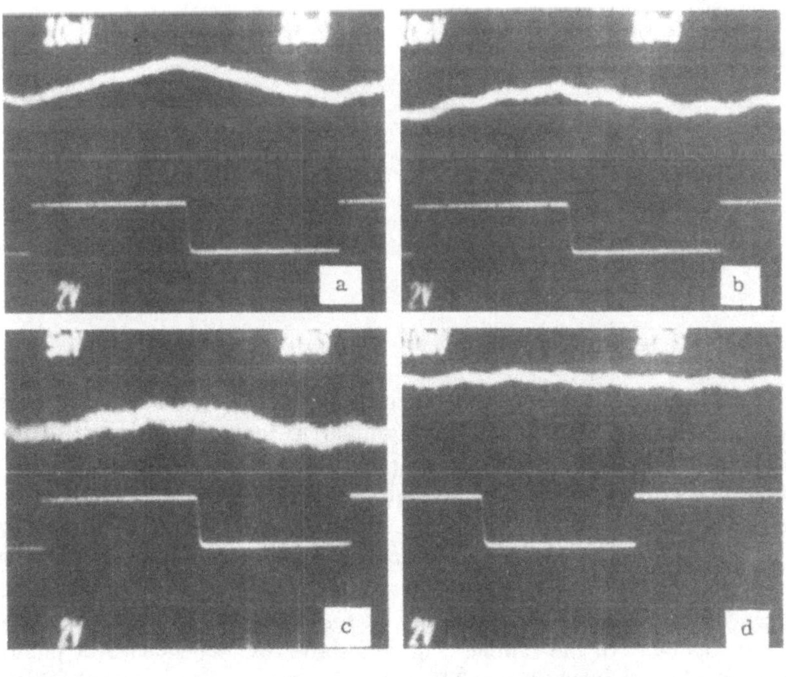

Fig. 7a–d. Photoacoustic signal of solid samples with monochromatic (1.56 Å) X-ray. **a** copper, R.C. 140.8 mA, **b** lead, R.C. 133.8 mA, **c** paper, R.C. 67.0 mA, **d** no sample, R.C. 108.0 mA. The samples are the same as in Fig. 5. The chopping pulse signals were in phase

Table 1. Photoacoustic signal intensities of solid materials

Sample (thickness)	White X-ray		1.56 Å X-ray	
	PA Signal/R.C. (mV/mA)	PA Signal/R.C. (μV/mA)	Mass Abs. Coeff[a] (μ) × 10^{-2} ($cm^2 \cdot g^{-1}$)	I/I_0
Copper (50 μm)	2.50	13.4 ± 0.3	0.551	0.0856
Aluminium (100 μm)	1.50	7.8 ± 0.4	0.527	0.243
Lead (400 μm)	1.02	6.8 ± 0.5	2.41	0
Glass (150 μm)	1.43	6.7 ± 0.7	—	—
Paper (100 μm)	0.69	2.7 ± 0.9	—	—
No Sample	0.63	2.2 ± 0.6	(0.083	0.998)[a]

R.C.; Ring Current of Stage Ring (mA), I/I_0; Transmittance
[a] Calculated for air (20.93 v/v% O_2, 78.10 v/v% N_2) (2 mm thick)

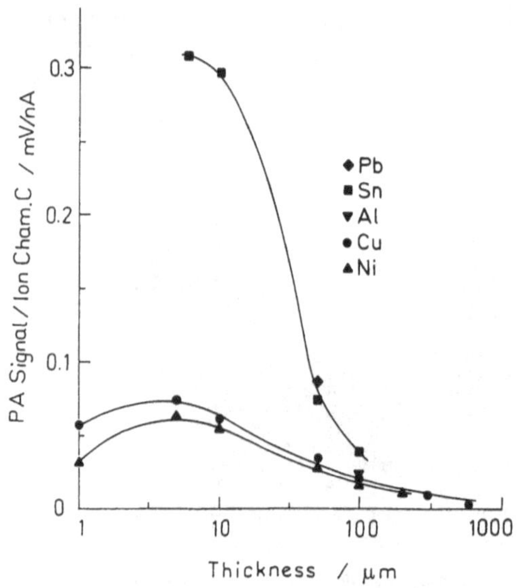

Fig. 8. The dependence of photoacoustic signal on sample thickness. Chopping frequency 10 Hz

the study. Table 1 summarizes the signal intensity per unit ring current. Although the mass absorption coefficient of copper is similar to that of aluminium at this X-ray energy, the difference in density gave quite different transmittance (I/I_0) for X-rays. The total absorbed beam and thus the photoacoustic signal are higher for copper than those for aluminium. One unexpected result was on lead sample. Mass absorption coefficient of lead is 5 times higher than copper and its transmittance is almost zero, however the photoacoustic signal was about two times smaller than that of copper. Other factors which explain this inconsistency should be heat capacity, and/or heat conductivity of the sample. Figure 8 shows the dependence of photoacoustic signal on the sample thickness. The signal shows maximum where the thickness of metal samples are close to the X-ray absorption depth. The heat diffusivity is high for metal samples, thus, the thicker the sample, the more heat dissipates into the sample of high heat capacity. On the other hand, organic materials which is optically transparent and of low heat diffusivity gave the signal maximum almost in the region of the thermal diffusion length.

The mechanism of heat generation is not clear at this stage. Final step of heat generation should be lattice vibration and the early step seems to be X-ray induced ionization, in-elastic scattering and the stopping of photoelectron.

4 X-ray Photoacoustic Effect of Gas Phase [9]

The gas is another phase which is detectable by microphonic photoacoustic method. The focused beam (at Beam Line 15A (PF)) at wavelength of 1.56 Å was used for this experiment. Photoacoustic cell for gas phase measurement is shown in Fig. 9. The optical path was 10 cm ($\times$6 mm Φ with beryllium windows (18 mm Φ, 0.5 mm

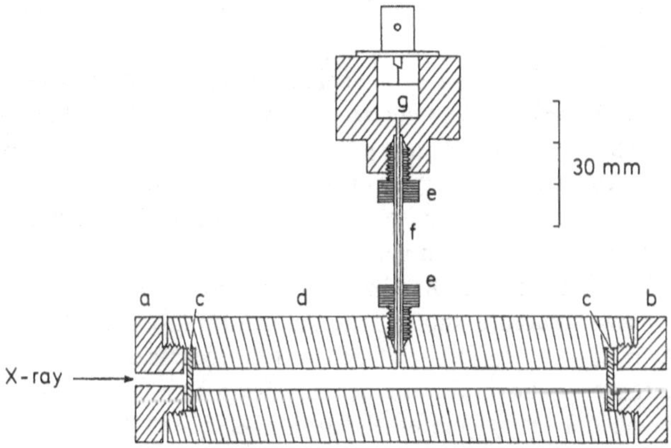

Fig. 9. Cross-section view of a photoacoustic cell for gas phase

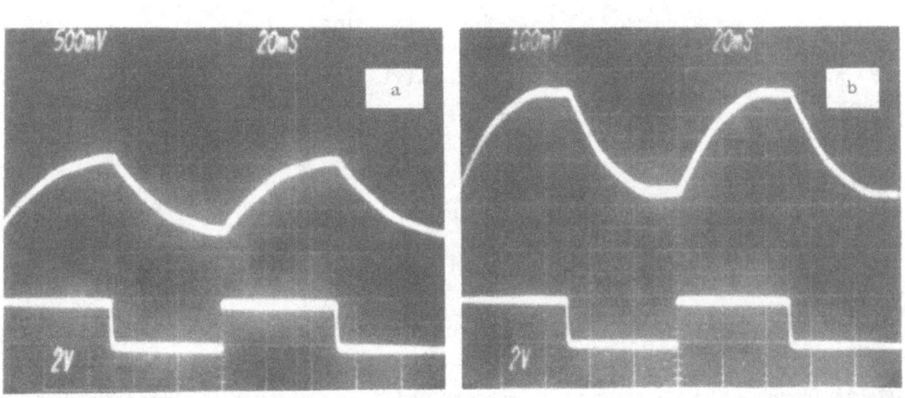

Fig. 10a and b. Photoacoustic signal shape of **a** xenon and **b** argon gas by 1.56 Å X-ray. Chopping frequency 10 Hz. X-ray chopping pulse is in-phase

thick) at both ends. The microphone was set at the end of the stainless microtube which was connected to the center of cell body. Argon and xenon gas and air were measured. Other photoacoustic apparatuses were the same as previously noted.

Figure 10 shows some typical signal shapes of X-ray photoacoustic effect seen for the gases. The rising-up time of the signal after the starting of irradiation of X-ray is usually faster than that for solid phase. It shows that the gas has lower heat diffusivity and less heat capacity than those for the solids. The mechanism of the X-ray photoacoustic phenomena for gas phase is still under speculation and should be 1) energy dissipation process into the heat energy and 2) ionization. In contrast to the solid phase, ionization process should be rather direct contribution for the photoacoustic effect. Various gases and various states of samples are under investigation by this method.

5 Photoacoustic Extended X-ray Absorption Fine Structure [5)]

Spectroscopic study was performed to reveal what kind of information is included in this heat generation.

The monochromatic X-ray was obtained by silicon (111) channel cut double crystal using white X-ray (at Beam Line 4A (PF)). The ion chambers were set at the both side of the photoacoustic cell, in order to compare the spectrum of photoacoustic X-ray absorption spectroscopy (PAXAS) with usual absorption spectrum, simultaneously. The chopper at chopping frequency of 10 Hz was set at the up-stream of these detectors. Copper foil (5 μm thick) was used as a sample.

Figure 11 shows the PAXAS spectrum and the absorption spectrum of the copper sample. Quite corresponding fine structure shows that the information of EXAFS is also included in the PAXAS spectrum. The heat generation process also reflects the EXAFS. The only difference is the monotonous increasing trend of PAXAS signal intensity along with the photon energy increase. This is also seen in the previous

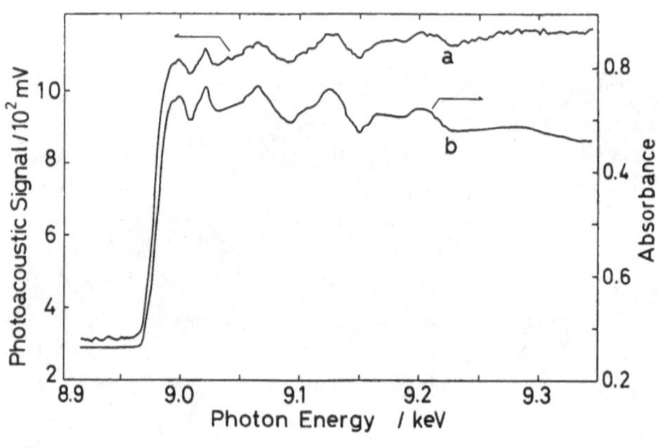

Fig. 11. Photoacoustic X-ray absorption spectrum and X-ray absorption spectrum for copper foil (5 μm thick). Photoacoustic signal is normalized by ion chamber current. Chopping frequency 10 Hz. Ring current 145–142 mA

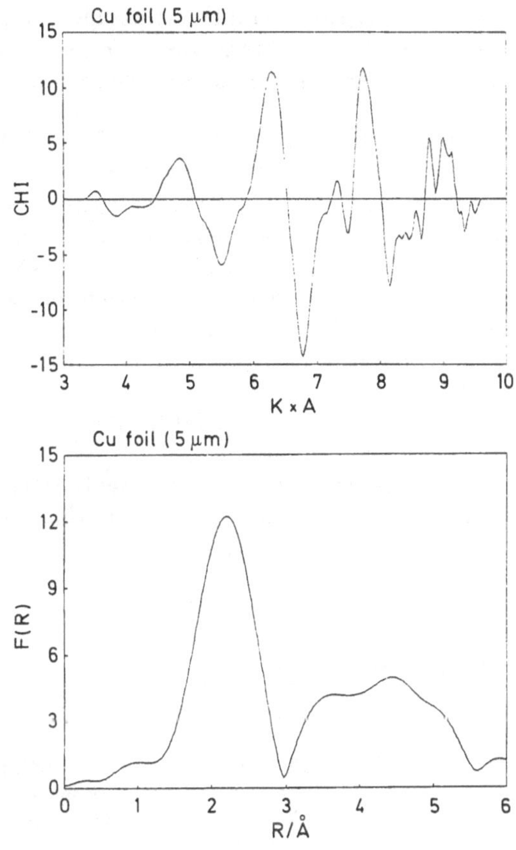

Fig. 12. Fourier transform of photoacoustic EXAFS spectrum in Fig. 11

rough PAXAS spectrum [6] and seems that the amount of heat depends also on the absolute photon energy. The more the photon energy is, the higher the PAXAS signal is. Figure 12 shows the Fourier transform of these data. Face center cubic lattice of copper gives the nearest neighbor atomic distance of 2.6 Å and the second and third nearest neighbor atomic distances of 3.6 and 4.4 Å. The peaks were seen in these region. The shoulder in the PA-EXAFS result (lower) around the 5 Å seems to be the effect of included noise in χ spectrum (upper).

Photoacoustic method can be applied any types of shapes and states of the sample and also to non-destructive depth-profiling. Thus this finding expands the methodology of EXAFS much wider than before. High temperature super conductor, biological samples and other varieties can be studied.

6 Depth-Profiling by Phase Analysis of Photoacoustic X-ray Absorption Spectroscopy [10]

As pointed out in the previous section, one of unique application of photoacoustic spectroscopy is depth-profiling. The heat which was generated under the surface, propagates to the surface. The time which is needed for propagation to the surface can be detected as the phase-lag. Phase analysis together with the amplitude of the signal can reveal the subsurface structure without destructive treatment of samples, and X-ray edge absorption is quite suitable for this purpose because of the specificity and discontinuity of absorption profile.

Nickel-plated copper foil (10 μm thick) was chosen as a model sample. The setting of apparatus is the same as that of photoacoustic EXAFS. The calibration of phase-lag using Nickel foils of various thickness was studied by separate experiment. One side of nickel foils which was painted in black was glued on the glass plate (1 mm thick) with optically transparent epoxy resin. From the bottom of the glass plate, He—Ne laser beam was irradiated in order to generate the heat wave at the bottom of the attached nickel foil. The heat propagates onto the surface and generates photoacoustic signal with certain phase-lag (at 27.6 Hz) which depend on the thickness of the foils.

Figure 13 shows the calibration curve of phase-lag for the nickel foils of various thickness. This result suggests the possibility of estimating the thickness of plated materials on the metal. Figure 14 shows the PAXAS spectrum of nickel-plated copper. In the photoacoustic amplitude spectrum, EXAFS of copper at the subsurface of sample was still detected clearly. This means that PAXAS method can be applied

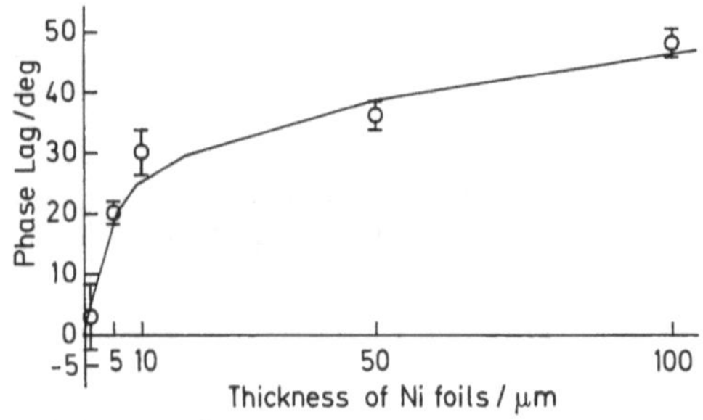

Fig. 13. Calibration curve of phase-lag for the heat wave propagation in nickel foil. Chopping frequency 27.6 Hz. See text for other experimental detail

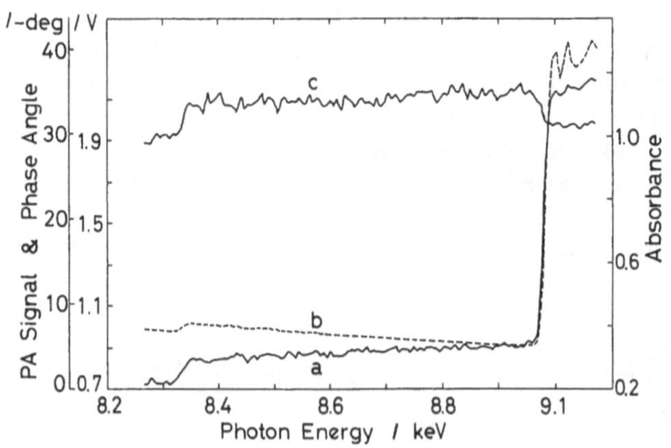

Fig. 14. Photoacoustic X-ray absorption spectra of (a) signal amplitude and (c) phase, and (b) absorption spectrum for nickel-plated copper foil (10 μm). Chopping frequency 10 Hz. Ring current 175 to 167 mA

even for the EXAFS analysis of subsurface materials. The phase spectrum shows sudden change at the K-edge points of nickel and copper and its lower shift means that the heat was generated mainly at copper area at subsurface. According to Fig. 13, the phase-lag is 4.6 ± 1 degree at 10 Hz. This value is converted to 7.6 ± 1.7 degree at 27.6 Hz [11] which is equal to be 0.5–2 μm of nickel-plate. This value was consistent with the data of 0.65 μm of nickel-plate layer which was estimated by the weight analysis.

7 Imaging Analysis by Photoacoustic X-ray Absorption Spectroscopy [12]

Photoacoustic imaging method is useful for the analysis of localization of various components in two-dimension [1], and photoacoustic method has one more advantage that it can be applied to the non-destructive depth-profiling. Thus, in future, 3-dimensional analysis should be possible by this method. For the development of this possibility, the model experiments were performed.

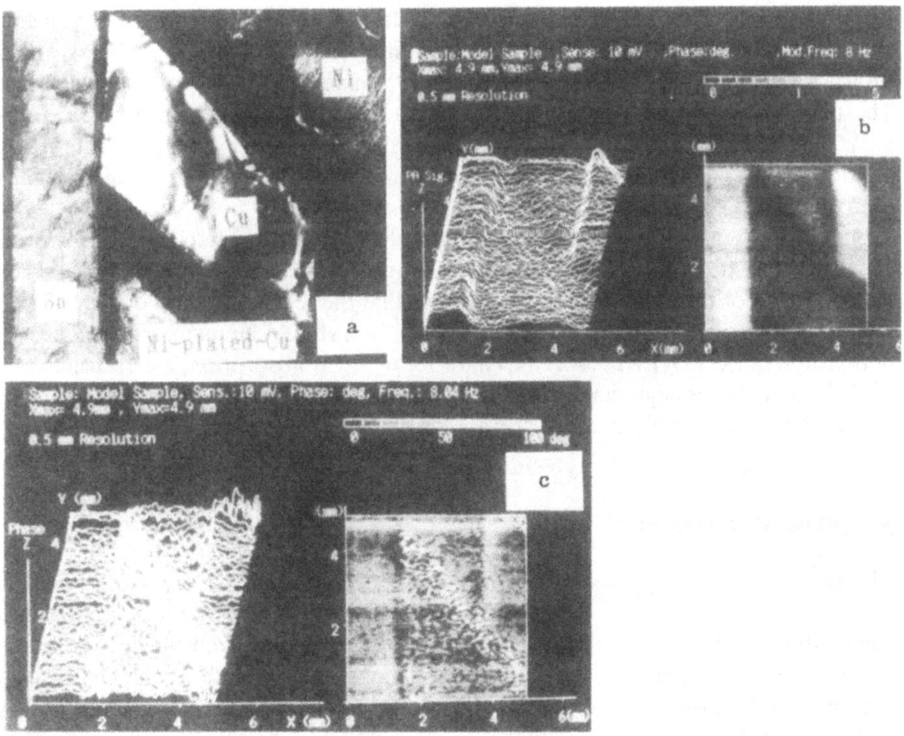

Fig. 15a–c. Photoacoustic X-ray absorption imaging using focused beam at 1.48 Å. **a** Photographic picture of the sample, **b** image of the signal amplitude, and **c** image of the phase. Chopping frequency 10 Hz

Focused X-ray beam (at Beam Line 15A) was used for this study. In order to improve the resolution, the focused beam (about 1 × 1.5 mm) was cut into 0.9–0.5 mm Φ by apertures which were set just in front of the cell. Modulation frequency was 10 Hz. Scanning X–Y stage which was originally developed for the laser microscopy [13] was set perpendicular to the surface of the iron-base table and scanning and data acquisition were controlled by PC-9801VM2 microcomputer (NEC Co. Ltd.) with the original program [13]. Various size and shape of metal foils were glued on the paper to have a model patterned sample.

Figure 15 shows the microscopic picture of the model sample. The photoacoustic imaging of this sample are Fig. 15(b) (photoacoustic signal amplitude) and (c) (photoacoustic phase). Since the wavelength of X-ray was 1.48 Å where the K-edge region of nickel locates, the image of signal amplitude (a) shows high signal region where tin and nickel foils locate, both of which have high absorption coefficient at this wavelength. Nickel-plated copper and copper foils gave small signal regions due to the small absorption coefficient or small amount of nickel. Even the same sample region, the difference in the amplitude can be seen which is mainly due to the difference in the thermal state of the foil. At the high amplitude region, the metal foil is not glued and thus stands without backing material (thermally floating). Consequently, this method also reveal not only the absorption profile but also the thermal circumstances. Phase image (c) was also displayed by the data processing program. Since the backing material (paper) gave small signal, phase image was noisy. In other regions, the phase values were a little different depending on the regions of samples which shows the difference in the depth of the point where the heat is mainly generated. The photoacoustic signal image together with the phase data can reveal the 3-dimensional image of materials and fundamental study is in progress. In soft X-ray region, K-edge detection for carbon by a pyroelectric thermal wave detector has been succeeded recently by Coufal et al. [14].

Now X-ray region has been opened to the photoacoustic spectroscopy. However, as seen in this text, the photoacoustic X-ray absorption spectroscopy is still in primitive stage. For the real applications, it seems that the sensitivity should be improved at least 10 times better than now or the photon flux should be increased by focusing or insertion devices. With the specific character of X-ray absorption, e.g. transparency of X-ray and abrupt edge shape absorption profile, this method seems to have hopeful future when the unique photoacoustic application can be conducted.

8 Acknowledgement

This study is partly supported by Grant-in-Aid for Scientific Research (No. 61870092, 62570968) from Ministry of Education, Science and Culture, Japan, by the Research Foundation for Pharmaceutical Sciences, and by the Mazda Foundation's Research Grant. The author greatly appreciates the collaboration and supports by Dr. Hiroshi Kawata, Prof. Masami Ando, Dr. Yoshiyuki Amemiya of Photon Factory at KEK, and collaboration with our research project members (28 persons) especially Prof. Hideo Imai (Fukuyama Univ.), Dr. Masaharu Hoshi (Hiroshima Univ.) and Dr. Mikio Kataoka (Tohoku Univ.) and also to the machine shop of Hiroshima Univ., Faculty of Science for their manufacturing of our apparatuses.

9 References

1a. Pao, Y.-H. (Ed.): Optoacoustic Spectroscopy and Detection, Academic Press, Inc., New York 1977

b. Rosencwaig, A.: Photoacoustic and Photoacoustic Spectroscopy, John Wiley and Sons, Inc., New York 1980

2. Bell, A. G.: Am. J. Sci. *20*, 305 (1880)

3. Masujima, T., Kawata, H., Amemiya, Y., Kamiya, N., Katsura, T., Iwamoto, T., Yoshida, H., Imai, H., Ando, M.: Photon Factory Activity Report *4*, 314 (1986)

4. Masujima, T., Kawata, H., Amemiya, Y., Kamiya, N., Katsura, T., Iwamoto, T., Yoshida, H., Imai, H., Ando, M.: Chemistry Letters *1987*, 973

5. Masujima, T., Kawata, H., Kataoka, M., Nomura, M., Kobayashi, K., Hoshi, M., Nagoshi, C., Uehara, S., Sano, T., Yoshida, H., Sakura, S., Imai, H., Ando, M.: Photon Factory Activity Report *5*, 139 (1987)

6. Masujima, T., Hoshi, M., Sugitani, Y., Sano, T., Nagoshi, C., Yoshida, H., Imai, H., Kawata, H., Amemiya, Y., Ando, M.: Photon Factory Activity Report *4*, 315 (1986)

7. Eyring, E. M., Komorowski, S. J., Masujima, T.: Analytical Instrumentation Handbook (Ed. Ewing, G.), Marcel Dekker, Inc., New York, in press

8. Hoshi, M., Masujima, T., Nagoshi, C., Sugitani, Y., Sano, T., Sawada, S., Kawata, H., Amemiya, Y., Ando, M.: Photon Factory Activity Report *4*, 313 (1986)

9. Masujima, T., Amemiya, Y., Kawata, H., Sakura, S., Hoshi, M., Uehara, S., Nagoshi, C., Sano, T., Yoshida, H., Imai, H., Ando, M.: Photon Factory Activity Report *5*, 348 (1987)

10. Masujima, T., Shiwaku, H., Yoshida, H., Imai, H., Kataoka, M., Hoshi, M., Sano, T., Ikeda, T., Makihara, H., Kawata, H., Amemiya, Y., Kobayashi, K., Ando, M.: Photon Factory Activity Report *5*, 140 (1987)

11. Adams, M. J., Kirkbright, G. F.: Analyst *102*, 678 (1977)

12. Masujima, T., Imai, H., Shiwaku, H., Kawata, H., Amemiya, Y., Iida, A., Hoshi, M., Nagoshi, C., Uehara, S., Kataoka, M., Danno, M., Ikeda, T., Makihara, H., Yoshida, H., Ando, M.: Photon Factory Activity Report *5*, 347 (1987)

13. Masujima, T., Munekane, Y., Kawai, C., Yoshida, H., Imai, H., Juing-Yi, L., Sato, Y.: Proc. of 5th International Topical Meeting on Photoacoustic and Photothermal Phenomena, Springer-Verlag, Heidelberg, Springer Series in Optical Sciences. vol. 58, pp. 19 (1987)

14. Coufal, H., Stohr, J., Baberschke, K.. Proc. of 5th International Topical Meeting on Photoacoustic and Photothermal Phenomena, Springer-Verlag, Heidelberg, Springer Series in Optical Sciences, vol. 58, pp. 25 (1987)

Author Index Volumes 101–147

Contents of Vols. 50–100 see Vol. 100
Author and Subject Index Vols. 26–50 see Vol. 50

The volume numbers are printed in italics

ISBN 3-05-500560-0

FSC
www.fsc.org
MIX
Papier | Fördert gute Waldnutzung
FSC® C083411

Zeitfracht Medien GmbH
Ferdinand-Jühlke-Straße 7
99095 Erfurt, Deutschland
produktsicherheit@kolibri360.de